Plants for Soil Regeneration: an Illustrated Guide

Plants for Soil Regeneration: an Illustrated Guide

by

Sally Pinhey and Margaret Tebbs

CABI

CABI is a trading name of CAB International

CABI
Nosworthy Way
Wallingford
Oxfordshire OX10 8DE
UK

Tel: +44 (0)1491 832111
Fax: +44 (0)1491 833508
E-mail: info@cabi.org
Website: www.cabi.org

CABI
WeWork
One Lincoln St
24th Floor
Boston, MA 02111
USA

Tel: +1 (617)682-9015
E-mail: cabi-nao@cabi.org

A catalogue record for this book is available from the British Library, London, UK.

Library of Congress Cataloging-in-Publication Data

Names: Pinhey, Sally, author. | Tebbs, Margaret Cecilia, author.
Title: Plants for soil regeneration : an illustrated guide / Sally Pinhey, Margaret Tebbs.
Description: Boston, MA, USA : CAB International, [2022] | Includes bibliographical references and index. | Summary: "This book gives farmers, growers and gardeners information and a visual guide to the plants which can be used to improve soil quality without the use of agrichemicals"-- Provided by publisher.
Identifiers: LCCN 2021057055 (print) | LCCN 2021057056 (ebook) | ISBN 9781789243604 (hardback) | ISBN 9781789243611 (ebook) | ISBN 9781789243628 (epub)
Subjects: LCSH: Soil restoration. | Soil science. | Soil management. | Soils--Quality.
Classification: LCC QH541.5.S6 P56 2022 (print) | LCC QH541.5.S6 (ebook) | DDC 577.5/7--dc23/eng/20211213
LC record available at https://lccn.loc.gov/2021057055
LC ebook record available at https://lccn.loc.gov/2021057056

References to Internet websites (URLs) were accurate at the time of writing.

ISBN-13: 9781789243604 (hardback)
9781789243611 (ePDF)
9781789243628 (ePub)

DOI: 10.1079/9781789243604.0000

Commissioning Editor: Ward Cooper
Editorial Assistant: Emma McCann
Production Editor: James Bishop

Typeset by SPi, Pondicherry, India
Printed and bound in the UK by Severn, Gloucester

Contents

Preface

It is neither fanciful nor an exaggeration to say that plants are the most essential elements of our world. From the earliest appearance of life on earth, they developed their own survival strategies, created soil, shaped atmospheric and soil conditions, and provided both homes and food for animals. We should respect, research and remember the amazing abilities of plants to adapt to adverse conditions, the flip side of which may be to alter soil conditions to both their and our advantage.

Growing food is nothing new. Traditional farming has been handed down through the generations for thousands of years, each generation doing what it believed best. Recent use of agrochemicals promised much in famine relief in the short term but created dangerous imbalances in elemental cycles.

The purpose of this book is to explore some of the ways in which plants may be used to fix or build cultivable earth, restore impoverished soil, maintain soil in good condition and decontaminate polluted soil. Many plants fall into more than one category. While not the apparent quick fix that chemical treatments offer, plant cover is cheaper, safe and natural. Plants do not need to be mined, and do not need vast factories or distribution networks. With a little help and judicious planting, they just get on with doing what they do best.

Of the many types of soil conditioning, we have selected six categories: soil stabilizers, green manures, herbal lays and cover crops, nitrogen fixers and phytoremediators. While reference may be made to tropical plants in common use, the plants illustrated in this book are temperate-climate plants. The list is not exhaustive. Plants are indexed with both their English and botanical names. The pages facing the illustrations address problems and solutions.

Agricultural expertise is a mix of well-known, tried and tested practices, and scientifically proven facts and effects. This book draws on all sources of information. Where something is unproven, the authors recommend it with caution. Many recent academic studies have focused on exploring traditional or folk beliefs and researching the scientific basis for them. This provides valuable information, but unfortunately such research is slow and expensive, and the range of subjects yet to be studied is unlimited. It would be very helpful to know, for instance, why a planting strategy usually works but sometimes does not. There are so many variables in plant and soil interactions that knowledge, care and diligence are the most important tools in using plants to restore damaged soils. This book is part of that movement towards sustainability.

Introduction by Dr Philip Cribb

When I was an undergraduate, the late Rachel Carson's *Silent Spring* first raised my awareness of the perils of meddling with nature without a full understanding of the consequences. At the time, the insecticide DDT (dichlorodiphenyltrichloroethane) was widely used around the world, particularly in the control of mosquitoes, the vectors of malaria, yellow fever and other debilitating and often fatal diseases. Its deleterious effects on birds and other organisms had passed unnoticed by the chemical companies and politicians who promoted its widespread use. Fortunately, her campaign led to its eventual banning. However, new perils to wildlife continue to emerge, the latest to receive widespread publicity being the catastrophic loss of insects, such as bees and butterflies, where neonicotinoids have been implicated.

Late in 1970, I landed in Peru on my first botanical expedition, calling first at the Lima International Potato Center. The driving force for the institute was the rapidly declining number of cultivated potato varieties being grown by the indigenous people in the Peruvian and Bolivian Andes where monocultures were being encouraged by multinational agrichemical businesses that has expanded into plant breeding. They were giving high-yielding cultivars free to the local farmers who forsook their indigenous varieties (over 2000 named cultivars at the time) with very mixed results. The new clones needed artificial fertilizers and pesticides to thrive, sold to the farmers by the same companies that provided the seed potatoes. In poverty-stricken communities, the lack of sustainability of such practices rapidly became apparent. Local farmers soon realized that the same pests and diseases attacking their new crop would have affected only a small proportion of their traditional crop and they would have been able to harvest enough each year to sustain their family life. Unfortunately, reversion to traditional practices was difficult because the old varieties were, by then, difficult or impossible to acquire. For me, it was an object lesson in how development needs to take into account a wide variety of issues that might not be immediately apparent. The politicians who pushed the groundnut scheme in East Africa in the 1960s also thought that they had backed a winner, despite scientists telling them that their environmental data were insufficient to guarantee success. The scheme failed when drought hit the region again and bankrupted local farmers, who had been led to think that they would gain a sustainable income for years to come.

The importance of taking a holistic view of the natural world has been reinforced for me over the years by my work on orchids at the Royal Botanic Gardens, Kew. Orchids live in a complex web of life, requiring a fungus to germinate their seeds and grow – and not any fungus but a particular one that can coexist with the orchid rather than destroy it. This is called a mycorrhizal fungus and it provides the orchid with nutrients to enable it to grow. Over the past few years, it has become increasingly apparent that mycorrhizal fungi in the soil live with most plants and are as vital to their survival and success as the climate, soil, ecology and pollinators. Mycorrhizal fungi form extensive networks in the environment often linking orchids and other plants with trees, bushes and decaying matter.

This has been termed the wood-wide web by mycologists, in allusion to its superficial similarity to the world-wide web that links the electronic world.

I have been fortunate to travel in many parts of the world in search of orchids. The rapidly increasing population and the increasing demand for resources, such as timber, land and mining, has placed formidable pressures on the natural world. As an outsider, it is easy to see solutions to the destruction of habitat but difficult to see how to promulgate these ideas to those who control the processes, usually politicians and big business managers and shareholders.

On the positive side, I have also seen many examples where individuals or dedicated groups of people have made a difference and saved habitats, species and livelihoods in the longer term. The first step is always to fully understand the environment and how it functions efficiently. On a small scale, this can apply to back gardens and allotments. On a larger scale, it can influence the management of farms, nature reserves and even newly built estates and developments. During the 1980s and 1990s, Kew established a network of universities, non-governmental organizations and small farmers that sought to provide quality botanical information to local farmers to enhance their lives in marginal environments. A thorough knowledge of the natural environment involves, at a basic level, the ability to name plants and animals. This book aims to assist users in selecting plants that can be grown to improve soil conditions in horticultural and agricultural environments in temperate zones. Sally Pinhey and Margaret Tebbs, our authors and artists, are among the finest botanical artists of their generation, associated with The Natural History Museum, Royal Botanic Gardens, Kew, and Chelsea Physic Garden where they have worked for many years. With sound ecological and botanical backgrounds, their art has been used to benefit science and the environment.

Their illustrations, all drawn from life, will make identification and selection easy. Their message that a healthy environment is essential to the long-term sustainability of farming is one that is becoming increasingly heard from campaigners and enlightened farmers, both here and abroad.

Phillip Cribb, Royal Botanic Gardens, Kew

Acknowledgements

Our grateful thanks to the following friends and colleagues for their help and encouragement: David Brown, Cheryl Cockburn, Philip Colfox, Tom Cope, Geraldine Darling, Ian Escott, Susan Flynn, Alison Groves, Nigel Hewish, Hilary Joyce, Sam Lane, Tim Parton, Roger Pinhey, Judith Saint, John Sheldon, Lydia Smith, Philip Sterling, Chris Stoate, Robin Walls and Ian Wilkinson, and especially to Tom Cope who saved us from our follies.

List of Plant Illustrations in Alphabetical Order of Latin Names

Artists: Sally Pinhey, Margaret Tebbs, Judith Saint, Susan Flynn, Ian Escott, Cheryl Cockburn and Geraldine Darling (indicated by their initials).

1 Soil Health

We generally think of the vigour of our plants as being the main test of soil health, and this is right. There are also many additional ways in which the health of soil can be tested. In the same way as animals alter the environment they live in by taking out nutrients and leaving waste matter, every growing plant will also alter the condition of the soil in some way. At the minimum, it will improve the organic content with its decaying form.

Starting from the very earliest plant life in the form of algae, mosses and lichens, photosynthesizing organisms survived on air, moisture, light and minerals in rock surfaces to grow and reproduce gradually. Over millennia, they created soil with their own decaying matter until it was deep enough to support larger plants that required foundations for their roots. This process can still be seen on any bare rock surfaces that have been newly created by quarrying or volcanic activity, for example. It is quicker, however, where there is already a surrounding community of plants or active construction. Soil is therefore composed of substantially dead plants with traces of minerals from rock substrate. Good soil consists of a mix of 40–45% inorganic matter, 5% living and dead organic matter, 25% water and 25% air. There is some flexibility in these percentages.

The depth of cultivable topsoil is also affected by its gradient and position with regard to the surrounding geology. Soil on steep slopes is prone to erosion and is thinner, while areas where sediment collects will differ from the bedrock, be deeper and consist of smaller particles. The activity of living organisms in the soil defines its quality. Living organisms affect soil structure by creating channels, with animals and microorganisms producing pores and crevices. Plant roots can penetrate into crevices to create friability, and strong, deep-rooted plants can break up compacted earth. Plant secretions promote the development of microorganisms around the roots. Leaves and other plant materials decompose and add to soil composition. Clearly, we do not know the entire story of soil dynamics, and there may be other approaches that our forebears, more conscious of natural cycles, understood better. Biodynamic systems (see Appendix) are demonstrably worthy of scientific study.

Microscopic organisms, so small that there would be millions in one teaspoon of soil, keep the soil alive with a variety of activities. They maintain the balance of life on earth by fixing gases and breaking down organic matter. These processes include breaking down bare rock into soil particles, cycling nutrients, transforming nutrients into different forms for plant uptake, helping the plants to absorb the nutrients, degrading toxins, both causing and preventing disease in plants, and both helping and hindering water penetration into the soil. A good balance of microbes in the soil is ideal for plant health, and this can be maintained by crop diversity, or at least small areas of monoculture together with crop rotation. Monoculture limits the type of microorganisms that can survive, and an imbalance permits the development of dominant pathogens.

As recently as 1996, glomalin was discovered by Sara Wright. It is a sticky substance that binds soil particles and sequesters carbon. It is thought to be produced mainly by mycorrhizal fungi that use it to protect their hyphae and strengthen the fragile fibres as

DOI: 10.1079/9781789243604.0001

they penetrate the soil and bridge air gaps, while converting nitrogen into a form that plants can assimilate. The network of glomalin from decaying hyphae binds the organic soil particles, improving both aeration and drainage. This process is improved by mulching but damaged by digging and ploughing.

Hermaphrodite, blind, deaf and supersensuous worm populations are both indicators and guardians of healthy soil. There are 29 earthworm species in the UK, of which there are four main types. Composter worms, usually bright red and striped, live in rotting vegetation and are excellent recyclers. Epigeic worms are reddish brown and live close to the soil surface, breaking down leaf litter. Endogeic worms live in deeper soil, are grey, pink, blue or green, and eat earth. Anecic worms have red or black heads with paler tails and live in burrows in the upper layers of the soil, pulling leaves down into their burrows from the surface. Following up Gilbert White's observations 100 years later, Darwin calculated that anecic worm castings add 5 cm of quality topsoil to the surface every 10 years. The worms do this by dragging dead leaves into their burrows, digesting them together with swallowed soil, and ejecting a fine calcareous and nitrogenous mix in their castings on the surface. Glands near the worm's gizzard add nitrogen and phosphates to their castings, while glands at the rear end add calcium, making the soil affected by their casts less acidic. Worm burrows also help to drain and aerate the soil, while waterlogged soils or those compacted by heavy machinery will have few or no worm populations.

Regular ploughing also chops up surface worms and exposes them to predators. Synthetic fertilizers and pesticides reduce the microbes that make the soil a healthy habitat for worms. Antiparasitic medicines given to grazing animals can reduce the number of microbes in the soil because of contaminated droppings.

A good mix of insects is an essential component of healthy soil. They are preyed on by spiders, which are a keystone species of the arthropods and a top predator, keeping the balance of pests and beneficial insects at an advantageous ratio.

Soil compaction is an increasing problem as farm machinery gets larger and heavier. Clay soils are particularly prone to this and are the hardest to restore. Even deep and strong-rooted plants like chicory, yarrow and sweet clover can sometimes fail to penetrate a clay pan. To some, the obvious solution may seem to plough it, but, paradoxically, not ploughing and leaving the soil organisms to do the work for you works better in the long run. Ploughing compounds the problem by extra heavy machinery passing over and chopping up the worms.

At the Allerton Project in Leicestershire, UK, where ploughing has been minimal over 25 years, the worm population has increased from single figures to an average of 800 per m^2. This is key to keeping clay soils profitably cultivable.

Knowing how long it takes for this precious earth covering to accumulate naturally, it follows that we should try to take good care of it, as it is the very foundation of our being. Farmers who have largely given up tilling say they can restore soil depth by up to 15 cm in as little as 2 years. The organic component of soil binds it, returns nutrients to the plant, stores moisture, makes soil friable and provides energy for soil microorganisms. Most soil microorganisms – bacteria, algae and fungi – are inactive when dry but become active once moisture is available. Good humus can hold water up to 75% of its volume without becoming waterlogged, rendering the land more drought resistant and preventing excessive runoff. Not all organic matter added to the soil is necessarily beneficial. In *Secrets of the Soil* (2004), Tomkins and Bird pointed out that cow manure is 25% microbes. Most of these are likely to be beneficial, but some may not be. Most microbes suppress diseases, but others may not. The effects will depend on the pathogens present and the amount and quality of the carbon:nitrogen ratio in the organic material, as well as moisture and temperature.

Insects expert Dr Jonathan Lundgren estimates that there are between 3000 and 15,000 species of potential pest insects, but for every pest species there are between 400 and 1700 species that are beneficial to humans. This is a good ratio, and is better undisturbed by indiscriminate insecticide use. On the whole, however, addition of organic material will suppress disease and increase the variety and activity of invertebrates and microorganisms in the soil. Careful selection of green manure plants is key to ensuring the best-quality organic additions. Soil damaged by repeated synthetic fertilizer applications can regenerate remarkably quickly if kept covered by a diverse mix of cover crops, which can also be grazed. The variety is essential, as plants with different requirements and different length roots coexist without competing and encourage a wider range of microorganisms. At Honeybourne Farm in the Cotswolds, UK, it was found that a sowing of mixed varieties produced a 50% increase in yield over a monoculture, known as the Darwin effect, having been noted by Darwin over 200 years ago.

We know that plants need water, light and nitrogen to grow. Nitrogen exists in the soil in two main forms: organic and inorganic. Crops cannot use organic nitrogen, but microbes in the soil convert it into inorganic nitrogen by a process called mineralization. Hungry plants, for instance those not fed with synthetic fertilizers, form stronger bonds with mycorrhizal fungi and symbiotic microbes, thereby enriching the soil with more diverse and vigorous microorganisms.

The speed at which organic matter breaks down in the soil is dependent on the carbon:nitrogen ratio of the matter. Soil composition will generally settle at a carbon:nitrogen ratio of around 12:1. The best ratio for decomposition is 24:1, which is the optimum for soil microbes, as too much carbon in the soil does not suit them. High-carbon plants such as rye with a ratio of 80:1 take a long time to break down, while low-carbon legumes provide a better ratio for rapid decomposition. This is an important consideration when choosing cover crops.

Plants also need trace elements in smaller quantities. The principal elements are phosphorous, magnesium, calcium, zinc, selenium and microscopic amounts of molybdenum. Molybdenum is essential for nitrogen fixation. In Australia, 28 g spread over 0.4 ha restored fertility for over 10 years. Trace minerals are absorbed by plants, largely through mycorrhizal fungi, becoming part of their structure and an essential element of human food. They are only required in minute quantities, but the band between a deficiency and toxic overload is also very small, and is measured in parts per million. This means that where naturally occurring elements are augmented by synthetic fertilizers, the potential for toxicity arises. A case in point is the use of phosphate fertilizers containing cadmium, or cadmium spread on farmland in contaminated sewage. The cadmium taken up by crops enters the food chain. Incidentally, smokers also inhale additional cadmium from cigarette smoke, tobacco being a plant that readily absorbs toxic minerals.

Food quality follows naturally from soil quality. Diversity of microbes in the soil enables plants to absorb micronutrients such as magnesium, calcium, zinc, sulphur and selenium, whereas synthetic fertilizers can only provide plants with nitrogen and phosphorous. The role that microorganisms play in the nitrogen cycle and in converting nutrients for plant use will be explained in Chapter 2 (this volume).

Nutrient data from 1930 to 1980 in the UK measured calcium, iron and potassium content in 20 different vegetables. Over these 50 years, the calcium content decreased by 19%, iron by 22% and potassium by 14%. Samples taken by government biochemists over a period of 51 years from 1940 to 1991 found potatoes had lost 47% of their copper, 45% of their iron and 35% of their calcium. Carrots, broccoli and tomatoes were in some respects worse. At the University of Texas, biochemists analysed the US Department of Agriculture nutritional data from 1950 to 1999 for 43 different vegetables and fruit, and recorded reliable information in loss of proteins, calcium, phosphorous, iron, vitamin B_2 and vitamin C over 50 years of increased use of artificial fertilizers. It now takes eight oranges to supply the same amount of vitamin C as was found in one orange 50 years ago. It is simple and cheap to measure the nutrient density of the food you grow with a Brix tester. Improvement in readings can be a guide to measuring the success of the planting on your soil. This has long been the tester of choice for the wine trade.

Bibliography

Brown, G. (2018) *Dirt to Soil*. Chelsea Green Publishing, Chelsea, Vermont.

Darwin, C. (1881) *The Formation of Vegetable Mould through the Action of Worms*. John Murray, London.

Rainbow, P.S. (2018) *Trace Metals in the Environment and Living Organisms*. Cambridge University Press, Cambridge.

Stoate, C., Leake, A., Jarvis, P. and Szczur, J. (2013) *Fields for the Future*. Game and Wildlife Conservation Trust, Fordingbridge, UK.

Tomkins, P. and Bird, C. (2004) *Secrets of the Soil*. Rupa Publications, New Delhi.

Tree, I. (2018) *Wilding*. Pan Macmillan, London.

White, G. ([1788] 1981) *The Natural History of Selborne*. Penguin, London.

Wright, S.F. and Nichols, K.A. (2002) Glomalin: hiding place for a third of the world's stored soil carbon. *Agricultural Research* magazine, pp. 4–7. Available at: https://agresearchmag.ars.usda.gov/2002/sep/soil/ (accessed 23 September 2021).

Dr Jonathan Lundgren is Director of the Ecdysis Foundation, and CEO for Blue Dasher Farm in South Dakota.

2 Nitrogen-fixing Plants

Nitrogen comprises 80% of the air we breathe. Nitrogen exists in many forms, both inorganic and organic, and is essential in an inorganic form in the air and soil for healthy plant crops. It is an essential part of all nucleic and amino acids required for cell division and growth. Plants that can fix nitrogen in the soil, in a symbiotic relationship with bacteria, are therefore especially valuable. Atmospheric nitrogen must be processed, or 'fixed', into a usable inorganic form to be taken up by plants. Between 5 and 10 billion kg of nitrogen $year^{-1}$ are fixed by lightning strikes, but most fixation is done by bacteria. Bacteria in a symbiotic relationship with plants have the enzyme nitrogenase, which combines gaseous nitrogen with hydrogen to produce ammonia, which is then converted by the bacteria into other compounds.

The relationship between the plant and the bacteria may be very specific, whereby the plant may only host one type of bacterium, or the possible combinations of hosts and bacteria may be more general. In the process of 'biological nitrogen fixation', soil bacteria respond to a chemical message from a nitrogen-starved plant root and fix on to the root hairs. They colonize the cells and produce a nodule in which the nitrogen from the air spaces in the soil is converted into ammonia with other organic compounds that the plant can use, giving it an advantage in poor soil, while also enriching the soil. There are two types of bacteria, symbiotic bacteria, which form nodules, and cyanobacteria, which are effectively algae, making their own nutrients by means of photosynthesis. Cyanobacteria fall more into the category of green manures, and are explored further in Chapter 4 (this volume).

Members of the genus *Azotobacter* are bacteria that fix nitrogen in the soil and may be used commercially to make biofertilizer. *Azotobacter* also synthesize some biologically active substances, such as the phytohormone auxin, which stimulates plant growth. This increases crop yields.

Azotobacter also contain antibiotics, which help the seedling to resist disease, and can help to rid the soil of heavy metals such as cadmium, lead and mercury. If the bacteria are not present in the soil, this cannot happen, so seed for commercial growers can be inoculated with the appropriate bacteria to encourage a good rate of nodule formation. This is a very complex interaction, currently the subject of much research. Nevertheless, experiments to date are promising, showing a future for biofertilizers and the reduction of fertilizer application costs and associated fertility loss. *Azotobacter* species have been found to fix atmospheric nitrogen, dissolve plant nutrients such as phosphates and stimulate plant growth by the production of plant growth hormones such as auxin, cytokinin and gibberellins. Bacteria can be added to the soil in a sterile and neutral carrier. Maize, mustard, rapeseed, sorghum and sugarcane, as well as tomatoes and lettuce, have produced enhanced yields under controlled conditions.

Root hairs absorb nitrate or ammonium from the soil that can be used by the plants. The nitrates absorbed are reduced first to nitrites and then to ammonia for conversion into amino acids, nucleic acids and chlorophyll. In plants that have a symbiotic relationship

DOI: 10.1079/9781789243604.0002

with rhizobia, some nitrogen is assimilated in the form of ammonia directly from the nodules. We now know that there is a more complex cycling of amino acids between rhizobia bacteroids and plants. The plant provides amino acids to the bacteroids so ammonia assimilation is unnecessary, and the bacteroids pass amino acids with the newly fixed nitrogen back to the plant, thus forming an interdependent relationship. While many animals, fungi and other organisms obtain nitrogen by ingestion of amino acids, nucleotides and other small organic molecules, other organisms including many bacteria can use inorganic compounds such as ammonium as their sole nitrogen source. The use of various nitrogen sources is carefully regulated in all organisms.

Legumes are the plant family best able to host symbiotic bacteria, and it is the conversion of fixed nitrogen into protein in the seed that makes them an important food source. The bacteria are non-toxic to humans and animals. Nitrogen-fixing bacteria only work in warmer weather. Once nitrogen is fixed in the earth, it must be captured. This can be done with ryegrass, which uses the nitrogen and then releases it back into the soil when it dies. The presence of *Brassica* spp. prevents nitrogen-fixing bacteria from forming.

The Nitrogen Cycle

Nitrogen comprises much of the earth's atmosphere and is a component of all living things. The nitrogen cycle (Fig. 2.1) describes the various forms through which nitrogen passes and the agents that cause the process of changes in it. Nitrogen is present in the

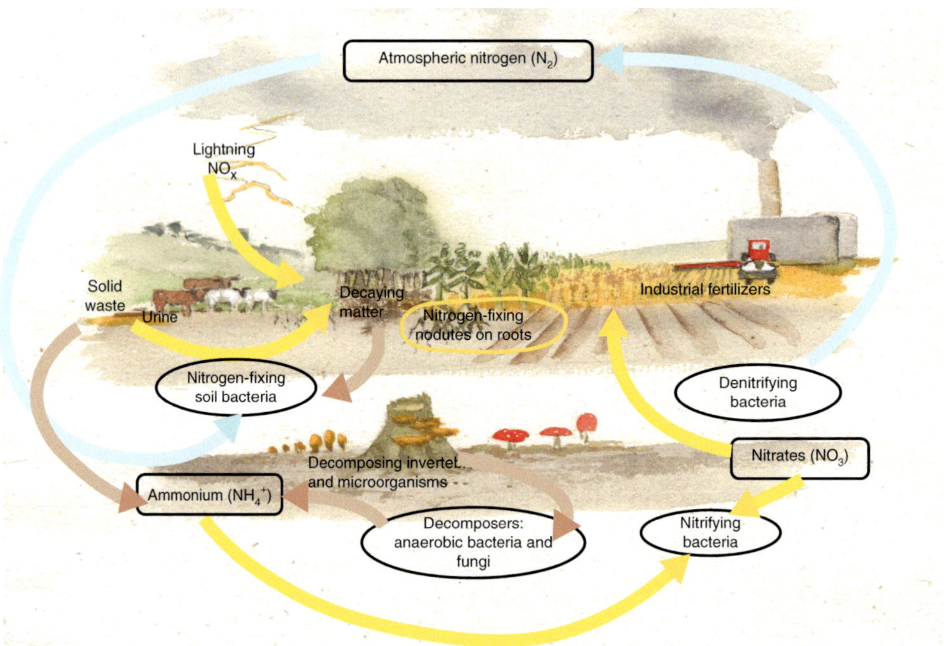

Fig. 2.1. The nitrogen cycle. Blue arrows indicate exchanges between atmospheric nitrogen and nitrogen-fixing bacteria in the soil. Brown arrows indicate the conversion of decaying matter into ammonium (NH_4^+). Yellow arrows show nitrates in the form NO_2, which is the form directly available to plants. Illustration by Sally Pinhey.

environment in a wide variety of chemical forms including organic nitrogen. As nitrogen is essential to plant growth, this description of the cycle takes the availability of nitrogen to plants as its central point. The various forms of nitrogen are usually referred to by the following chemical formulae:

- Inorganic atmospheric nitrogen gas is N_2.
- Ammonia is an inorganic gas and the form that plants can use, either dissolved as the ammonium ion (NH_4^+) or as ammonia gas (NH_3). It is common in animal waste.
- Nitrite is NO_2. It is an inorganic ion (a particle with a positive or negative electrical charge) that is very reactive and only occurs naturally at very low concentrations.
- Nitrate is NO_3. It is an inorganic ion and is water soluble. It may be present in high concentration in rivers, lakes and estuaries.
- Nitrous oxide is N_2O. Small amounts are present in the atmosphere. It is sometimes known as laughing gas, and becomes liquid under pressure. It has many commercial uses and so is generated by human activity. It is harmful to the ozone layer.
- Nitric oxide is NO. It is an organic gas, and can be generated by lightning and created by bacteria and plants. With a short-lived facility to penetrate cell walls, it is used as a signalling system within organisms.

Organic nitrogen may be in the form of living organisms or humus. The main processes through which nitrogen changes form are called fixation, ammonification, nitrification and denitrification.

Nitrogen fixation is the process by which atmospheric nitrogen is converted either by a natural or an industrial means to a form of nitrogen such as ammonia. In nature, most nitrogen is harvested from the atmosphere by microorganisms to form ammonia, nitrites, and nitrates that can be used by plants. A small amount is converted to usable forms by lightning in a process called atmospheric nitrogen fixation. Fixation done industrially makes artificial fertilizers.

3 Cover Crops, Green Manures and Herbal Leys

Bare soil is the enemy of soil health.

Niels Corfield (2019).

Soil that is left bare, especially over winter, risks damage by frost, heavy rain and strong winds. By using plants as cover crops, green manures and herbal leys, these problems can be averted, and the ground will be improved in terms of both structure and nutrient content. When the plants die, their remains will be drawn down by worms, resulting in deeper, more fibrous soil with better absorption of water. Moisture will also be held for longer in a drought. Pastures with cover crops are better able to absorb and retain heavy rain compared with bare, degraded land. The numbers of beetles and worms will increase, enriching the soil content by burrowing, eating and excreting after digesting organic particles. Microorganisms (bacteria, archaea and fungi; discussed in Chapter 4, this volume) that live in and maintain soil structure will thrive on a hearty mixture of carbon and nitrogen provided by the plants via photosynthesis. These organisms are particularly numerous around plant roots (the rhizosphere), where sloughed-off cells from the roots provide them with nutrients. As primary decomposers of organic matter, they fix nitrogen to help growing plants, suppress harmful chemicals (toxins), and benefit and stimulate plant growth. In addition, many bacteria dissolve phosphorus (essential for healthy plant growth), making it available for plants to use.

Bacteria benefit from the sugars secreted by plant roots. They do not thrive in dry conditions, or in acid, salty, compacted or poor soil, but can take some ground disturbance. Fungi do better in soil with minimal disturbance or in a no-dig system. Ploughing damages and breaks up the hyphae (root system), and fungicides will destroy beneficial as well as noxious species. Moist conditions with plenty of organic matter allow fungi to increase and benefit the soil. Many plants develop a beneficial relationship with fungi that increases the contact of their roots with the soil. Fungi penetrate the roots, sending out hyphae and absorbing water and nutrients that feed the plant. The hyphae are filamentous and able to reach into tiny spaces in the soil, inaccessible even to plant roots. Soil-dwelling nematodes are minute, non-segmented worms about 1 mm long. They live in the moisture surrounding soil particles and are crucial to maintaining soil fertility.

Research has also shown that some flying pests can be confused if the outlines of their food plant are disguised. For example, underplanting brassicas with legumes such as trefoil or vetch disguises the outline of the crop and seems to confuse and deter cabbage root fly. Slug predators such as frogs and beetles enjoy the cool, damp ground under a crop of green manure.

The nitrogen cycle (see Figure 2.1, Chapter 2, this volume), in which atmospheric nitrogen is converted into different organic compounds, is vital to sustain all living organisms. During the cycle, microorganisms in the soil process or 'fix' atmospheric nitrogen into ammonia, which plants need in order to grow. Other bacteria convert the ammonia into amino acids and proteins. Animals eat the plants directly or indirectly, and consume the protein. Nitrogen compounds return to the soil through animal waste, which is converted

DOI: 10.1079/9781789243604.0003

by bacteria back to nitrogen gas and returned to the atmosphere. A nitrogen molecule in the air consists of two atoms, which are held tightly together (N_2). To enable the absorption of nitrogen, these atoms must be separated. Bacteria (diazotrophs) in the soil are able to convert molecular nitrogen from the air to ammonia (NH_3), breaking the atoms apart. This is called nitrogen fixation.

Lightning and solar radiation also contain enough energy to separate the nitrogen molecules in the air. Once apart, they fall to earth with rainwater, and combine with soil minerals to form nitrates. However, the amount of nitrogen fixed in this way is negligible compared with the amount fixed by microorganisms in the soil. Nitrates and nitrites occur in igneous rock, and are widely used by the agrichemical industry as inorganic fertilizers.

Too much nitrogen applied to the land can cause problems by leaching into water courses and groundwater. Animal wastes and nitrogen-containing fertilizers increase concentrations of nitrates in the environment. According to the US Environmental Protection Agency (EPA), pollution caused by excess nitrogen and phosphorus in water is widespread and costly, and can cause significant health problems.

What Is the Difference?

Cover crops comprise a range of plant species, preferably a mix that includes legumes, brassicas and grasses. This mix provides the soil biology with carbon and nitrogen in the correct amounts needed for survival. These crops are usually sown on farmland in autumn, to overwinter and keep soil protected while not being used for other agricultural or horticultural purposes. This protection can ensure against soil erosion and weed competition, especially during periods of high rainfall. Cover crops are now one of the most common agricultural strategies for improving soil health, aiding water retention and preventing nitrogen pollution (runoff) in water. They provide the soil microorganisms with carbon, which is used for energy.

Green manures have an emphasis on adding organic matter and nitrogen to the soil, especially where there are no animals to add manure. Green manures are meant to be sown and harvested before maturity, and either dug in or, preferably, left on the surface of the ground to be drawn down by worms. They return valuable nutrients to the soil and improve structure. With their deep root systems, these plants gather nutrients from depths that ordinary vegetables rarely reach. Fast-growing green manures such as annual clovers (*Trifolium* spp.) will benefit the soil between intensive crops, boosting nitrogen content. They may be grown for a whole season, as can non-legumes such as mustard and phacelia (*Phacelia tanacetifoli*). Perennial leguminous plants such as lucerne (*Medicago sativa*) and sainfoin (*Onobrychis viciifolia*), which take longer to establish, add carbon and nitrogen over a longer term. Usually sown with grasses, and suitable as grazing or silage, green manure plants also protect against soil erosion. Intercrop sowing is a way of establishing green manures between a cash crop, such as cereals or brassicas. They do well together, and when the crop is harvested, the green manure continues to grow, keeping the soil protected. Green manure crops need to be carefully integrated into the crop rotation, and proper attention must be paid to their husbandry. The cost of seed can be high, and land may be tied up while they grow. Recent legislation aimed at reducing the environmental impact of farming has led to a renewed interest in these crops and the important part they can play in soil health.

Herbal leys comprise a seed mixture of fast-growing grasses, legumes and other herbaceous plants, which can aid livestock health, provide grazing and improve soil fertility. There may be a mixture of up to 17 species, depending on location and soil type (e.g. clay, sand, chalk). As part of a rotation, the mix will build up fertility. The plants capture carbon and nitrogen from the atmosphere and take it down through the roots where it sustains microbes in the soil. Farmers who use herbal leys in a rotation find that in time they need to spend less or even nothing on chemical fertilizers.

Species such as chicory (*Cichorium intybus*) and sainfoin contain tannins in their stems and leaves, which can reduce the presence of animal gastrointestinal nematodes. Tannin-rich plants are noted for their effect on internal parasites in ruminants. Condensed tannins are polyphenolic compounds, secondary metabolites of plants that are concerned with protection against inflammation, ultraviolet radiation or aggression by pathogens. In the last decade, there has been interest in these naturally occurring tannins as a way of avoiding the use of 'drenches' as a cure for parasitic infestations in livestock. These natural tannins act by reducing nematode viability and decreasing egg hatching, larval development and larval stages. They also act indirectly, increasing host resistance by binding to dietary protein and providing protection in the rumen. Studies investigating the effects of plants bioactive against gastrointestinal nematodes have generally involved grazing sheep, with most published work focusing on forage legumes such as sweet vetch (*Hedysarum coronarium*), sainfoin, trefoils (*Lotus* spp.), sericea (*Lespedeza cuneata*), as well as chicory in the Asteraceae. Tannin chemistry is now being linked with agronomy (the study of soil structure), plant breeding, animal nutrition and parasitology.

Inevitably, there is a certain amount of overlapping between cover crops, green manures and herbal leys. What is clear is that when any plant decomposes, the nutrients captured become available to subsequent crops. Deep-rooted cover and green manure crops that are left to grow for several months or more can acquire minerals and other nutrients at levels that most vegetable roots would never reach. These nutrients rise up through the roots and vascular systems, and when the crop dies, the debris enriches the upper levels of soil. Fast-growing plants are sown to add carbon and nitrogen to the soil as well as fibrous matter. Cover crops are either producers of soil nitrogen (legumes) or 'scavengers', such as cereal and annual ryegrass, and brassicas such as oilseed radish (*Raphanus sativus*). 'Scavenging' means being able to trap or take up soil nitrates that would otherwise be lost to leaching. Some legumes can also scavenge residual nitrogen but are slower to establish and therefore less practical.

Plants from the family Leguminosae, such as the clovers (*Trifolium* and *Medicago* spp.) are able to absorb nitrogen from the atmosphere and fix it in nodules on the roots. These nodules degrade once the plant has died. This process feeds the countless soil bacteria, thus creating a fertile growing medium known as humus. Non-leguminous plants can 'hold' or 'forage' nitrogen in the soil, making it available to the next crop. Members of the Brassicaceae are particularly good at this. The cut plant debris can remain on the soil surface. It is not necessary to plough or dig, and is better for the soil fauna if left alone. Severe soil disturbance can alter the delicate balance of microorganisms and other minute creatures that inhabit the substrate, and may stimulate weed seeds to germinate. Once a cover crop is mature, it is cut, or finished, with a 'no-till crimper roller'. The trick is to time the growth so that the plants are tall enough to be flattened but before they seed. This bulky cover suppresses weeds and will allow the next crop to come through after drilling. In agricultural land, crops can be planted directly into the surface detritus, using a no-till drill, which scrapes a small area on the surface and deposits a seed. Some farmers prefer to kill their cover crops with chemicals, to remove competition to cash crops. However, research has shown that the benefits to the soil biome are considerably diminished by the use of herbicides.

Cover crops, green manures and herbal leys may be annuals or perennials, and should be sown when the soil has warmed up after winter, or in the autumn while weather conditions are still good. Broad-leaved plants such as phacelia (family Hydrophyllaceae) and brassicas (family Brassicaceae) will cut out light available to less acceptable, often ground-hugging weeds, and will aid drainage and 'forage' nitrogen to make it available for the next cash crop. It should be noted that members of the Brassicaceae are susceptible to a stunting disease called club root and are best rotated. Generally, all roots should be left in the soil to rot down, adding to the soil structure and helping to prevent erosion. Once the soil content starts to improve, worms and insects become more numerous, and microbes multiply.

In summary, cover crops, green manures and herbal leys fertilize the soil in virtually the same way that fallow periods do, and in the long term can maintain fertility. While the constant use of chemical fertilizers degrades soil structure and depletes it of microbes and worms, these plants can restore the balance.

Humus

When leaves, stems, twigs and other material fall to the ground, litter is formed. Leaf litter is more than just a pile of fallen leaves. These and other tree components also act as mulch and sustenance, and serve as shelter to insects and microorganisms. When insects, birds and animals die, their remains are added to the litter. Over time, all of this matter decomposes, breaking down into important soil nutrients. The crumbly brown or black substance that remains after most of the litter has decomposed is called humus.

Earthworms live and multiply in this rich mixture, and help mix humus with minerals in the soil. Humus contains many useful nutrients for healthy soil. One of the most important is nitrogen, a key nutrient for plants. Humus adds fertility and structure to the soil, and helps to grow healthy plants.

Soil Health and Regenerative Agriculture

Although humans have been producing crops for millennia, it is only from the mid-20th century that agrichemicals have taken a huge hold on crop production. The new artificial fertilizers were considered miracles of growth, and it seemed that they would be able to feed the world. However, although crops have been bountiful for some years, the sustained and regular use of artificial fertilizers, pesticides and herbicides is now taking its toll. Soil in some areas has become thin, dusty, degraded and lacking in nutrients, worms and microorganisms. Ploughing or digging every season has damaged the soil structure even more, and erosion is commonplace. All farmers would like to stop erosion, but many will grudgingly sacrifice their soils rather than change practices that have been long established. Droughts add to the pressure, soil is blown away and a vicious circle is established of more soil depletion. The appalling dust storms of the 1930s in the mid-western USA were caused by excessive ploughing up of the prairie. The government of the time implemented some changes to land use to keep the soil in place. However, the effects of drought and the overuse of underground water for irrigation are again becoming evident. Heavy applications of chemical fertilizers have led to an imbalance in the nitrogen cycle and consequently to surface and groundwater pollution. Excess nitrogen fertilizer leaching into fresh water, as well as into marine ecosystems, has caused eutrophication. This is where too much nitrogen in the water results in dense growths of surface algae (an algal bloom), which block out the light. The outcome is the death of many organisms, resulting in little or no aquatic life.

Pesticides kill beneficial insects as well as harmful ones, thus reducing predators and pollinators. The loss of biodiversity has led to a reduction in nutrients, needing more applications of synthetic fertilizer. Weeds thrive on nitrogen and multiply, leading to a greater use of herbicides. Some weeds have become resistant to the heavy use of herbicides such as glyphosate. Many herbicides used today are 'chelators', small molecules that bind tightly to metals such as zinc, manganese, magnesium, iron and copper, preventing their uptake by plants. These essential elements are needed to keep plants healthy and to ward off fungal diseases. The application of more and more fungicides is detrimental to both soil biology and pollinators. Recent studies have shown that fungicides, once thought to have no ill effects on bees, weaken their health and make them susceptible to mite infestation. Scientists and corporate executives have to acknowledge that these compounds are having harmful effects, impacting on disease resistance and animal health. Although the chelating properties are well known, this additional hazard was not adequately considered in the regulatory risk assessment. A lack of nutrients available to plants in soil also makes them more susceptible to pest attack. An increase in pest infestation leads to increased use of pesticides. The majority of pesticides are not pest specific, which means that many beneficial insects are killed, including species such as bees that are needed to pollinate crops. Almost all fruit and vegetables grown on conventional farms today are sprayed with copious amounts of insecticides. These are not only absorbed into our bodies but also into those of other creatures.

Winds of Change

In the 21st century, many farmers are changing their cultivation practices. Depressed by the high cost of agrichemicals, and with declining harvests and soil erosion due to flooding or drought, they have been sowing cover plants along with cash crops on their land. When ploughing is reduced or stopped (no-till farming), and artificial fertilizers are reduced, worms and arthropods return. The soil depth and friability improve, thanks to cutting and leaving plant residues on the surface. Worms draw the organic matter down, and soil fungi and bacteria start multiplying. Along with the use of carefully controlled animal grazing and their manure, land is being restored to vitality. Not all of these farmers consider themselves organic growers; they are simply trying to reduce their reliance on chemicals and adapt to nature's rhythm again. They want to make a decent living, and soil health is a proven key to growing successful crops. Diversity (not being confined to only one crop, known as monoculture), plus the use of cover plants to attract pollinating and predating insects and birds, helps prevent pest attacks. Deep topsoil packed with nutrients can absorb more water and therefore is not so susceptible to flooding or erosion. Yields and plant health will improve.

Animal manure as part of farm management is important in the cycle of maintaining crop fertility. Many arable farmers are able to borrow some stock to graze their land (mobbing). Sheep or cattle will eat the cover plants, treading the stems in and leaving manure. Well-made compost is probably the best of all soil additives but can require a lot of labour to produce enough. Where possible, a mixture of cover plants, animal dung and compost is the most practical option.

Commonly used grass cover crops include cereals (rye, wheat, barley, oats) and forage grasses such as ryegrass (*Lolium perenne*). A problem common to all grasses is that if the crop is grown to maturity, it can take all available nitrogen from the next crop. This is because of the high carbon and low nitrogen content in grasses nearing maturity. This can be solved by growing grass with legumes and other broad-leaved plants in a mixed system. Cover crops such as brassicas are useful for trapping or 'scavenging' nutrients, especially nitrogen, left over from a previous crop. They tend to have extensive root systems, which can grow deeply into the soil, aerating and aiding drainage (e.g. forage or oilseed radish (*R. sativus*) or field radish (*R. raphanistrum*)) and reducing erosion. In addition, they can produce large amounts of residue, thus adding organic matter to the soil. The large leaves help suppress weed germination and growth by eliminating light reaching the ground.

Practising no-till methods, where seed is sown directly into a small cut in the ground, does not interfere with the integrity of the soil. Reduction of or stopping the use of chemical fertilizers, herbicides and pesticides can save a huge amount of money. However, it can take several seasons for good fertile soil to build up. Harvests will reflect this, and some farmers will compromise, using small amounts of artificial fertilizers and herbicides to begin with. Keeping the land covered at all times by plants protects it from harm. Whether the soil is heavy and clay-like or light and sandy, the penetrating root systems of cover plants will improve structure and hold moisture. Using a mix of plants, especially ones with large leaves, will promote photosynthesis and thus add more nutrients to the soil. Weeds will be suppressed by these larger plants, which can cut out the light.

Chemical and Nutritional Composition of Brassicaceae

Brassicas contain isothiocyanates, thiocyanates and nitrites as major sources of their distinctive flavour, pungent sulphur-like aroma and bitter taste. Brussel sprouts are rich in vitamin A and ascorbic acid, and contain appreciable amounts of riboflavin, niacin, calcium and iron. Cabbage is rich in vitamin C and high in minerals and proteins. Chinese cabbage provides 25 mg of vitamin C, 240 IU equivalents of vitamin A, and moderate amounts of calcium, iron, phosphorus and potassium per 100 g of raw product. The cold-hardy varieties possess higher soluble sugar content. Kale (*Brassica oleracea*) is a rich source of vitamin C, carotinoids and calcium. It is believed that brassicas have been cultivated for several thousand years, but their history as a domesticated plant is not clear before Greek and Roman times, when it was already a well-established vegetable. The Greek philosopher Theophrastus (371–287 BCE) mentions three kinds of *rhaphanos* – curly-leaved, smooth-leaved and a wild type. The ancients believed that cabbages grown near grapes would impart their sulphurous flavour to the wine. Brassicas have been bred over time into many varieties, including cabbage, broccoli, cauliflower, Brussels sprouts, collards and kale. Cover crop mustard (*Brassica nigra*) seeds contain a compound called sinigrin, which is a glucosinolate (a natural component of certain pungent plants such as mustard, horseradish and cabbage). Whenever sinigrin is crushed, the enzyme myrosinase is released, creating a mustard oil. When the seeds are crushed and mixed with cold water, an oil results that can actually cause burning or blistering when it comes in contact with the skin.

Farming Without Man-made Chemicals

Farming based on the management of soil life and ecology to optimize forage and crop productivity involves feeding the soil, not the plants. This holistic approach involves the recycling of nutrients from manures, absorption of minerals and improved soil structure from cover crops. The emphasis on soil health and structure distinguishes chemical-free farming from conventional, which often relies on artificial fertilizers, pesticides and herbicides. There is a thriving movement towards organic farming in parts of the USA. For example, on a small farm in Georgia, USA, the land was plagued by Bermuda grass (*Cynodon dactylon*), a tough drought-tolerant grass, that was choking out cash crops. This grass spreads rapidly by stolons and rhizomes, quickly infesting an area. Neighbouring farmers used Roundup, burnt off the grass or tilled the ground to get rid of it. However, these methods left dry, barren soil that needed to be replenished. Two crop strips were set up at the small farm – one ploughed and sprayed before sowing the cash crop, and the other untilled and sowed with cover crops such as buckwheat, millet and peas, along with the cash crop. The first strip was left with dry, thin soil and a reduced crop. The second strip was left with healthy rich soil, and the crop yields were good. The farmer now only uses cover crops, and is convinced that the soil is improved.

Plant Species used in Cover Crops, Green Manures and Herbal Leys

Much research has been done on a wide variety of plant species to discover which are most beneficial for nitrogen or carbon. The following list deals with the most commonly used plants.

Asteraceae (Compositae)

Chicory (*C. intybus*) is tall with beautiful blue flowers and penetrating roots. The taproot is long and thin, with numerous secondary roots that can break through compacted soil, helping to drain and aerate it. It contains tannins in the stems and leaves.

Brassicaceae (Cruciferae)

The family Brassicaceae is also known as Cruciferae, meaning 'cross-bearing' in reference to the regular four-petalled flowers. Brassicas used as cover crops include mustard, rapeseed and forage radish. They are generally used as winter or rotational cover crops in vegetable and specialty crop production, such as potatoes and tree fruits. Rape (*Brassica napas*) is killed by harsh winter conditions but is grown as a winter crop in milder parts.

Oilseed, Daikon and field radish have gained much interest because of their fast growth, which allows significant uptake of nutrients. These radishes develop a large starchy taproot 3–6 cm in diameter and 30 cm or more long that can break through compacted layers, allowing deeper rooting by the next crop. Radishes should decompose by spring, but leave the soil in a friable condition and improve water infiltration and storage. They also ease root penetration and development by the following crop. Canola (*Brassica napus*) and other *Brassica* crops may function as biofumigants, suppressing soil pests, especially root pathogens and plant-parasitic nematodes. They are excellent rotation crops, but pest management results are inconsistent. White mustard (*Sinapis alba*) is a rapidly growing plant that is easily incorporated at the end of the season. Not fully hardy, it dies back in cold weather. It does not need to be dug in, just left on the surface to break down. Plants in the Brassicaceae used as cover crops provide a good source of nitrogen for the following crops. Members of this family do not develop mycorrhizal fungal associations.

Hydrophyllaceae

Phacelia (*P. tanacetifolia*) is a very hardy plant, used as a cover crop or green manure for overwintering. As a fast-growing hardy annual, it can be sown from March until September, and has a powerful scent. The vibrant, blue-purple flowers are very attractive to bees and hoverflies. It can germinate at low temperatures, and its dense ferny foliage can smother weeds. The extensive root system improves soil structure and prevents leaching. It grows very firmly and is difficult to dislodge. Later sowings can overwinter, and have been seen growing in fields in January.

Leguminosae

A group of bacteria (*Rhizobia* spp.) live inside the specialist nodules that form on legume roots and provide nitrogen that can be used by the plants. The bacteria benefit from the sugars secreted by the roots. Many legumes contain tannins. Some studies indicate that using cover crops, especially legumes, between main crops helps maintain high levels of spores and promotes good mycorrhizal development in the next crop.

Lucerne (*M. sativa*) can grow for several years, depending on factors affecting survival of the tap roots. It will not thrive on waterlogged or heavily compacted soils. Lucerne is slightly more susceptible to downy mildew and attack from sitona weevil than red clover. It can also suffer from stem nematode infection and *Verticillium* wilt.

Sweet clover (*Melilotus officinalis*) has an erect open habit and so is not a great weed suppressant. It can produce good biomass once established, and has the potential to fix nitrogen. It can survive over winter as a tap root.

Sainfoin (*O. viciifolia*) can be slow to get going, but once established produces enough cover to compete against weeds. It contains tannins in its stems and leaves.

Fenugreek (*Trigonella foenum-graecum*) has a long growing season and thus considerable bulk production may be achieved. It is fast to establish, producing a quick boost to soil fertility in just a few months.

Clover (*Trifolium* spp.) provides an excellent long-term cover crop. Weed seeds in the disturbed ground will germinate with the clover, but if all are cut back, the clover will regrow and smother any regrowth of weeds. The nodules provide valuable nitrogen. There are several species and varieties of clover:

- Red clover (*T. pratense*) is one of the most tried and tested green manures for short- to medium-term leys, and is popular with organic farmers. It can grow rapidly, with reasonably good persistence for up to 3 years. Red clover silage has very good protein levels.
- White clover (*T. repens*) has large-leaved varieties, which are the most productive and best suited to cutting or light grazing. The leaves die back in winter to an underground mass of stolons, which are tolerant of all but the most severe frosts. White clover is slow to establish, but increases in size later in the season and in subsequent years, offering good weed control. Much breeding effort has gone into varieties of white clover, and many have good tolerance to common pests and diseases such as *Sclerotinia* infection and stem nematodes.
- Berseem or Egyptian clover (*T. alexandrinum*) is an annual clover that provides fast-growing biomass and nitrogen fixation but needs warm soils to germinate.
- Persian clover (*T. resupinatum*) is another annual that is sown more in Europe. It can grow rapidly but needs warmth to germinate. It has a sweet fragrance.

Field beans (*Vicia faba*) are a smaller variety of broad beans. Although used as a green manure in a garden situation, particularly on heavy soil, the cost of seed can be prohibitive on a field scale. In addition, field beans are not completely frost hardy and may be particularly susceptible to the disease 'chocolate spot'.

Vetch (*Vicia sativa*) is extremely competitive against weeds, forming an impressive cover with its sprawling habit. When incorporated, the residues also have an allelopathic effect, inhibiting germination of other seeds. This effect persists for around six weeks, and an adequate interval should be left if drilling direct sown crops after a crop of vetch.

Polygonaceae

Buckwheat (*Fagopyrum esculentum*) is a summer annual that is easily killed by frost. Sowing needs to be left until May in temperate climates, but once established, it will grow better than many other cover crops on low-fertility or acidic soils, rather than in nitrogen-rich soil. It grows rapidly and completes its life cycle quickly, taking around 6 weeks from planting in a warm soil until the early flowering stage. Buckwheat can grow more than 60 cm tall in the month following planting. It can be used to suppress weeds following an early spring vegetable crop, and has also been reported to suppress important root pathogens, including *Thielaviopsis* and *Rhizoctonia* spp. It is possible to grow more than one crop of buckwheat per year in many regions. It can seed rapidly and become a problem unless mowed before maturity. Buckwheat is a highly nutritious staple food suited to a short- or late-season crop. It secretes chemicals that attract phosphorous-producing organisms, and will restore soils that have phosphorous lockout. It needs little care if sown in a fine seedbed once frosts have passed, and is sometimes used as a green manure, as a plant for erosion control, or as wildlife cover and feed. As a cover crop, it establishes quickly, suppressing summer weeds and providing good biomass at up to 3000 t acre^{-1}, preparing the ground for a following crop. As it dies quickly in frost, it requires no topping, and can be left on the soil surface. Its flowers attract pollinators, which greatly increase the yield, and the nectar produces a dark-coloured honey. As it hosts aphid predators, it can be interplanted as a companion to crops that suffer from aphid infestations.

Grasses Used as Cover Crops, Green Manures and Leys

Italian or Westerwold ryegrass (*Lolium multiflorum*) is suitable for building soil structure in orchards, vineyards and other cropland to enhance water-holding capacity. It can reduce soil splash on fruit crops, decrease disease and enhance forage quality. Italian ryegrass has good water and nitrogen holding properties, and makes an excellent cover crop. When sown in autumn, ryegrass may reduce the severity of root rot caused by a fungal disease, *Rhizoctonia*. Residues are slow to decompose, with a carbon:nitrogen ratio of about 80:1. It can be grown with vetch and later grazed, after which another crop can be seeded into the trampled soil. It is a good choice for mixing with high-value crops.

Oat (*Avena sativa*) is an annual plant that can tolerate higher rainfall and is suitable for northern climates. It provides a naturally killed mulch the following spring and may help with weed suppression. Left on the surface, oat can provide cover, and dead stems help trap snow and conserve moisture.

Sudan grass (*Sorghum* × *drummondii*) and *Sorghum*–Sudan hybrids are fast summer annuals that produce a lot of growth in a short time. Because of their vigorous nature, they are good at suppressing weeds. However, they need a minimum ground temperature of 18–20°C to germinate. Sudan grass has been reported to suppress plant-parasitic nematodes and possibly other unwanted organisms. Sudan grass is especially helpful for loosening compacted soil. It can also be used as a livestock forage and so can do double duty in a cropping system with one or more grazings. It and its relatives contain a glucoside that releases prussic acid after cutting. This breaks down in a few weeks, after which the residue is safe to use.

Winter rye (*Secale cereale*) grows rapidly in the early spring. It is very hardy and easy to establish. It is suitable for planting in, and will grow and overwinter readily. Decomposing residue of winter rye has been shown to chemically suppress germination of weed seeds.

Pesticides and Herbicides

A brief mention of these chemicals is necessary, because of their detrimental effects on soil inhabitants, as well as on both beneficial and unwelcome insects. Further details can be found in the Appendix.

Glyphosate is absorbed through foliage and minimally through roots, and goes to the growing points (meristems). It inhibits a plant enzyme involved in the synthesis of three aromatic amino acids: tyrosine, tryptophan and phenylalanine. It is therefore effective only on actively growing plants. It is used to quickly 'kill off' green manures before planting cash crops.

Neonicotinoids, which are nerve agents, have been shown to cause a wide range of damage to individual bees, such as memory loss and a reduction of queen numbers. Evidence has strengthened recently of neonicotinoid damage to colonies of bees, and has revealed a huge reduction in all flying insects. It is mostly banned in Europe and some other countries in order to protect pollinators.

Aminopyralid and clopyralid are hormone-based herbicides, usually sprayed on meadows grazed by horses. They mimic plant auxins, and persist in manure for several years. The effect is to stunt growth in broad-leaved vegetables, such as broad beans, lettuces (*Lactuca* spp.) and tomatoes (*Solanum lycopersicum*). Vegetable growers have been alarmed by their persistence, especially as they may take several years to disperse. They were banned for a few years, but have recently been reinstated. Grasses are not affected.

Asulam is a herbicide used to control bracken, a toxic fern that grows on acid ground.

Bibliography

AgMRC (2018) Buckwheat. Agricultural Marketing Resource Center, Ames, Iowa. Available at: www.agmrc.org/commodities-products/grains-oilseeds/buckwheat (accessed 27 September 2021).

Canola Council of Canada (2021) History of canola seed development. Canola Council of Canada, Winnipeg, MB, Canada. Available at: www.canolacouncil.org/canola-encyclopedia/history-of-canola-seed-development/ (accessed 11 October 2021).

Corfield, N. (2019) What is soil health and why should farmers and growers care about it? *Vidacycle*, online. Available at: https://soils.vidacycle.com/soil-health/what-is-soil-health-and-why-should-farmers-and-growers-care-about-it/ (accessed 1 December 2021).

EPA (2021) *Groundwater and Drinking Water: National Primary Drinking Water Regulations*. US Environmental Protection Agency, Washington, DC. Available at: www.epa.gov/ground-water-and-drinking-water/national-primary-drinking-water-regulations (accessed 27 September 2021).

Justice Pesticides (2021) Neonicotinoids in courts. Justice Pesticides, France. Available at: https://justicepesticides.org/en/2021/03/09/les-neonicotinoides-au-tribunal/ (accessed 27 September 2021).

Kim, N., Zabaloy, M.C., Guan, K. and Villamila, M.B. (2020) Do cover crops benefit soil microbiome? A meta-analysis of current research. *Soil Biology and Biochemistry* 142: 107701.

Mueller-Harvey I., Bee, G., Dohme-Meier, F., Hoste, H., Karonen, M. *et al.* (2019) Benefits of condensed tannins in forage legumes fed to ruminants. *Crop Science* 59, 861–885.

Ota, R., Ohkubo, Y., Yamashita, Y., Ogawa-Ohnishi, M. and Matsubayashi, Y. (2020) Shoot-to-root mobile CEPD-like 2 integrates shoot nitrogen status to systemically regulate nitrate uptake in *Arabidopsis*. *Nature Communications* 11: 641.

Rayns, F. and Rosenfeld, A. (2010) *Green Manures – Species Selection*. Garden Organic Factsheet 25/10. Available at: www.organicresearchcentre.com/manage/authincludes/article_uploads/iota/technical-leaflets/green-manures-species-selection.pdf (accessed 27 September 2021).

Rothamsted Research (2021) *The History of Rothamsted Research.* Rothamsted Research, Harpenden, UK. Available at: www.rothamsted.ac.uk/history-and-heritage (accessed 27 September 2021).

Sands, B. and Wall, R. (2018) Sustained parasiticide use in cattle farming affects dung beetle functional assemblages. *Agriculture Ecosystems and Environment* 265, 226–235.

Taylor, J. (2021) Inspired by family history, Georgia farmer shows the way for organic best practices. *Eco Farming Daily*, online. Available at: www.ecofarmingdaily.com/farm-management/inspired-by-family-history-georgia-farmer-shows-the-way-for-organic-best-practices/(accessed 27 September 2021).

Wills, R. (2020) A guide to mob grazing livestock. *Farmers Weekly*, online. Available at: www.fwi.co.uk/livestock/grassland-management/a-guide-to-mob-grazing-livestock (accessed 27 September 2021).

4 Bacteria and Other Microorganisms

> Producing more food with fewer resources may seem too good to be true, but the world's farmers have trillions of potential partners that can help achieve that ambitious goal. Those partners are microbes.
>
> Reid, A. and Greene, S.E. (2012).

Current conventional farming concentrates on eliminating the harmful microbes that cause disease, and in the process kills all the beneficial ones as well. If the beneficial ones can be encouraged and outweigh the harmful ones, then great savings can be made on the costs of pesticides and fertilizers, while helping to reverse environmental and climate damage. In the complex world of competition and cooperation in soil life, nothing is simple; yet from what is already evident, certain planting strategies are known to be effective. For example, chicory (*Cichorium intybus*) sown with white clover (*Trifolium repens*) benefits from the additional nitrates in the soil around nodules on the clover roots. However, if ryegrass (*Lolium* spp.) is also sown in the mix, it responds to nitrates provided by the clover bacteria and helps retain the nitrogen in the system. This also reduces the severity of *Rhizoctonia* root rot, a fungal disease affecting seedlings in a humid atmosphere. Inoculation is another strategy sometimes advised for species such as sweet clover (*Melilotus* spp.), fenugreek (*Trigonella foenum-graecum*) and lucerne (*Medicago sativa*), as they form associations with a less common bacterium (*Rhizobium meliloti*), which may not be present in the soil. Maize sown with an inoculation of *Azotobacter* showed increased yield, and a practice is now growing up of using biofertilizers, which are effectively a soil additive of cultured microbes tailored to the soil structure and anticipated crop.

Microscopic organisms – so small that millions would fit in one teaspoon of soil – keep soil alive with a variety of activities, maintaining the balance of life on Earth by fixing nitrogen, breaking down organic matter and preparing it for plant uptake. Atmospheric nitrogen is the most abundant gas in the atmosphere, but plants cannot use it as they do not have the necessary enzyme to fix it and convert it into proteins. Bacteria, cyanobacteria and other minute organisms fix nitrogen and convert it into ammonia (NH_3). This process can provide 65% of the nitrogen used in agriculture. Other processes include breaking down bare rock into soil particles, cycling nutrients, transforming nutrients into different forms, helping plants to absorb the nutrients, degrading toxins (both causing and preventing disease in plants), and helping or hindering water penetration into the soil. A good balance of microbes in the soil is essential for plant health, and this can be maintained by crop diversity, or by small areas of monoculture together with crop rotation. Monoculture limits the types of microorganisms that can survive, creating an imbalance that permits the development of dominant pathogens.

Bacteria

Bacteria were the first living organisms to develop on Earth over 4 billion years ago, followed by cyanobacteria with the ability to photosynthesize after a further half a billion years. Cyanobacteria live in aquatic conditions and are beneficial in paddy fields.

DOI: 10.1079/9781789243604.0004

Consisting of 60% nitrogen, cyanobacteria are a valuable nitrogen pool in the soil by performing a critical function in the nitrogen cycle (see Chapter 2, this volume). Some species convert animal waste into usable nitrogen for plants. Others are responsible for nitrification and denitrification. Nitrifying bacteria convert the organic nitrogen in waste into forms that plants can assimilate. Denitrifying bacteria live in airless conditions such as waterlogged soil using nitrate instead of oxygen to 'breathe' and in the process returning nitrogen gas to the atmosphere. Where denitrifying bacteria are abundant, they will reduce the soil fertility; examples are *Thiobacillus denitrificans*, *Micrococcus denitrificans*, *Paracoccus denitrificans* and *Pseudomonas* spp.

Bacteria have the ability to reproduce rapidly, doubling their number every few hours under optimum conditions. They thrive in damp soil but can also survive extended periods of drought in the form of a dormant cyst. Most bacteria are immobile and have an external digestive system that releases enzymes into the soil. These enzymes can be used by other organisms. It is estimated that soil could contain up to 3 million species of bacteria, of which roughly 5% have been named. Humus and plant cover on the soil help to maintain favourable conditions for many beneficial species.

Some bacteria have the ability to form symbiotic associations with plants. The plants provide the bacteria with organic chemicals, mainly carbohydrates manufactured by photosynthesis, while the bacteria provide mineral nutrients to the plant. Among the nitrogen-fixing bacteria that form this type of relationship with plants are *Azospirillum* spp. in the *Rhizobium* group, which form nodules on the roots of leguminous plants, clearly visible to the naked eye, and *Azotobacter*, a genus of nitrifying bacteria that fix atmospheric nitrogen making it available to the roots of plants in the surrounding soil. Their activity is inhibited where soil is already high in nitrogen, and plants are less likely to form a relationship with other organisms.

Non-symbiotic microorganisms also benefit from secretions from plants roots, but it is the bacteria that form a symbiotic relationship with plants that are of particular interest to farmers. Research into soil microbes is continually making new discoveries, but these symbiotic bacteria have been known since 1888, when Martinus Beijerinck discovered what it was that made the visible nodules on the roots of legumes – it was not a disease but a small nitrogen factory. In return for carbohydrates made by photosynthesis in the plant, the rhizobia bacteria convert organic nitrogen that is unusable to the plant into inorganic nitrogen compounds that can be used in the plant to make amino acids and proteins. It is no coincidence that plants that have this symbiotic relationship with bacteria are those with the highest protein content and among the most important world food sources, such as beans and groundnuts. Mankind is therefore also a symbiont of the legume, protecting, breeding and promoting it with the widespread practice known as farming. Although it had long been noted that these legumes added fertility to soil, it has only recently been found that plants use most of the nitrogen for their flowers and seeds, leaving only about 3% in the soil. Legumes grown purely for soil enrichment should be turned in or grazed before they flower, when the nitrogen content is at its highest. A few non-legume plants have the capacity to form a relationship with bacteria. Alders, bayberries, sweetfern and casuarinas all have the symbiont *Frankia*, which is an actinobacterium.

Microbiologists are increasingly finding evidence of bacteria, fungi, nematodes and viruses forming symbiotic relationships with plants, enabling them to withstand the stresses of disease, toxins and climate. Linda Kinkel at the University of Minnesota, Minneapolis, has discovered that some plants actively encourage *Streptomyces*, a genus of bacteria that produce antibiotics to control certain pathogens. Green manures can alter the balance of microbes to increase this activity, an approach proven to reduce soil-borne diseases. These discoveries are key to producing enough food for the projected population, and while the effects of the microbial community can be replicated and proven in a controlled environment, applications to farm management are more complex. Some ready-packaged microbe inoculants, which include *Azospirillum*, *Azotobacter*, *Bacillus* and *Paenibacillus* spp. and *Trichoderma harzianum* are designed to bioremediate soils contaminated with pesticides and herbicides. Known as 'plant growth-promoting rhizobacteria', they can mitigate some of the damage done by chemical fertilizers.

Fungi

Fungi were traditionally considered to be plants and have only in recent times been acknowledged as a kingdom in their own right, with more in common with animals than plants. The kingdom includes yeasts, smuts, dry rot, mildews and moulds, which are only a small percentage of the whole kingdom comprising around 144,000 known organisms. Fungi differ from plants in that they have no chlorophyll and cannot make their own carbohydrates. They obtain their nutrients from chemicals in nature and their main structural element is closer to that of insects than plants. The familiar toadstools that appear above ground are only a small part of their life cycle, which for the most part takes place in dark, moist spaces, linked by thread-like growths called hyphae. These hyphae are collectively called mycelium, and can extend many miles and penetrate between the cells of dead and living organisms, in some cases in a mutually beneficial relationship. The process of extracting nutrients from dead organic matter is the most important part of decay and recycling. In dry areas where fungi do not thrive, dead organic matter builds up and cannot be used by plants. While penetrating plant roots for the sugars produced by the plant, some fungi also convert chemicals in the soil into a form available to the plant, for example phosphates. This is known as a mycorrhizal symbiosis. Crop rotation improves mycorrhizal fungal activity, especially in the pasture phase, because grasses and clovers associate with the fungi. Phosphate inhibits this symbiotic relationship, and so less artificial fertilizer is needed during the pasture phase.

Work done by Jarmo Holopainen and Suzanne Simard has revealed a complex and dense world of cooperation and competition in the soil based on communication via fungal hyphae.

Other microorganisms

Besides bacteria and fungi, other soil residents are micro-arthropods, protozoans, nematodes, earthworms, insects and viruses. They live on organic matter or on smaller prey and perform vital recycling of nutrients. Nematodes are minute translucent 'worms' with thousands of species varying from 0.5 to 50 mm long. Some are harmful and damage crops, such as tomatoes, potatoes, peppers, lettuce, maize, carrots, onions, rye and lucerne. Others eat weevils, clearwing borers, cutworms and webworms, cinch bugs, sod webworms and white grubs. The latter are the nematodes that indirectly break down compost, and can be bought and spread as a pest deterrent.

A good mix of insects is also an essential component of healthy soil and all form part of the complex soil community of prey and predators. Some, together with spiders, are keystone species and top predators maintaining a balance of pests and beneficial insects. It is difficult to anticipate the exact processes in the interaction of plants and microorganisms in the complex soil community, but, in general, a close association between plant roots and microorganisms enhances plant nutrient uptake where the organisms can access nutrients not otherwise available to plants and convert them into assimilable forms.

Bibliography

Barr, D.J.S. (1992) Evolution and kingdoms of organisms from the perspective of a mycologist. *Mycologia* 84, 1–11.

Barth, B. (2014) Microbes will feed the world, or why real farmers grow soil, not crops. *Modern Farmer*, online. Available at: https://modernfarmer.com/2014/04/microbes-will-feed-world-real-farmers-grow-soil-crops/ (accessed 28 September 2021).

Bulluck, L.R. III, Brosius, M., Evanylo, G.K. and Bristaino, J.B. (2002) Organic and synthetic fertility amendments influence soil microbial, physical and chemical properties on organic and conventional farms. *Applied Soil Ecology* 19, 147–160.

Elad, Y., Chet, I. and Katan, J. (1980) *Trichoderma harzianum*: a biocontrol agent effective against *Sclerotium rolfsii* and *Rhizocionia solani*. *Phytopathology* 70, 119–121.

Encyclopedia Britannica Online. Available at: www.britannica.com/ (accessed 28 September 2021).

Ferris, J. (2011) The secret life of plants. *New Scientist* 209, 46–49.

Gothandapani, S., Sekar, S. and Padaria, J. (2017) *Azotobacter chroococcum*: utilization and potential use for agricultural crop production: an overview. *International Journal of Advanced Research in Biological Sciences* 4, 35–42.

Kinkel, L. (2011) A coevolutionary framework for managing disease-suppressive soils. *Annual Review of Phytopathology* 49, 47–67.

Rayns, F. and Rosenfeld, A. (2010). *Green Manures – Species Selection*. Garden Organic Factsheet 25/10. Available at: www.organicresearchcentre.com/manage/authincludes/article_uploads/iota/technical-leaflets/green-manures-species-selection.pdf (accessed 27 September 2021).

Reid, A. and Greene, S.E. (2012) *How Microbes Can Help Feed the World*. Report on an American Academy of Microbiology Colloquium Washington, DC. American Society for Microbiology, Washington, DC.

Jarmo Holopainen works at the University of Eastern Finland, Finland, and Suzanne Simard in the Department of Forest and Conservation Sciences at the University of British Columbia, Canada.

5 The Role of Fungi

Mycorrhizae

'Mycorrhiza', a word coined by Albert Bernhard Frank in 1885.

Fungi are a vital part of the mixture of microorganisms found in healthy soil. All vascular plants associate with soil fungi and bacteria. The benefits are enhanced nutrient access and stress tolerance. Fungal associations between plant roots and beneficial fungi are known as mycorrhizae (meaning 'fungus' and 'root'), and form a beneficial or symbiotic relationship with plants growing in the soil. Mycorrhizal fungi also facilitate plant interactions with other soil microbes. These include pathogens, and bacteria that produce vitamins and protect against attack.

Mycorrhizal fungi have fascinated and baffled researchers for years. Starting at the end of the 19th century, researchers in the traditional fields of mycology and botany became aware of plant–fungal symbioses. One hundred and fifty years later, discussion and research continues to uncover amazing facts about these extraordinary organisms, as well as questions about their ecology and evolution. There are at least 6000 named species of soil fungi, and their structure and functions are found in all kinds of ecosystems. These ancient organisms, thought to be involved in the original colonization of the land, are both simple and complex. This group of fungi forms associations with host representatives from the whole plant kingdom. In this beneficial partnership, the fungi are supplied with nutrients, while the plant is able to access vital minerals and water, with some protection from pests, disease and soil pollution. There are many species of mycorrhizal fungi, found in all climates and soil types.

When fungal spores germinate, minute thread-like structures called hyphae emerge and spread through the soil, seeking out plant roots to invade. The hyphae form a wide web-like network that links plants underground and are able to penetrate the smallest cracks and fissures in the soil. They are able to absorb moisture and essential elements such as phosphorus, manganese, copper and zinc, and make these available to plants. Phosphorus particularly is difficult for plants to absorb without the actions of fungal hyphae.

Phosphorus enables plants to grow strongly and fast, and the survival of plant seedlings may also be up to five times greater if they have mycorrhizae to help them collect nutrients and water from the soil. Fungal hyphae also exude a sticky substance called glomalin, which can bind soil particles and elements together, creating air pockets and space for water to pass through. Plant roots provide energy from photosynthesis in the form of sugars to the mycorrhizae, and both organisms benefit. Mycorrhizae may also give some protection to plants from infections. Roots with plentiful mycorrhizae are better able to resist fungal diseases, parasitic nematodes, drought, salinity and heavy-metal toxicity. Mycorrhizal associations benefit the nitrogen-fixing bacteria *Azotobacter*, which also produce growth-stimulating chemicals. Mycorrhizal fungi can often form multiple associations with different plants. The fungal hyphae of several different kinds of fungi may resemble a giant underground network connecting plants in a habitat. It is possible

DOI: 10.1079/9781789243604.0005

that this enables different species of plants to exchange nutrients via the fungal hyphae. Access to this underground network of hyphae has the potential to improve seedling survival. Where mycorrhizae are absent from soil or present only as isolated spores, plant growth will be considerably reduced. In theory, plants can communicate with each other through underground mycorrhizal networks that interconnect the roots of multiple plants. In this way, defence against pathogens may also be communicated.

Soil cultivation can disrupt the actions of microorganisms, including mycorrhizae, and fertilizers can do more damage. The heavy use of phosphorus fertilizers in cultivated land is thought to suppress mycorrhizal activity, and fungicides will cause further damage.

Tilling the soil destroys the network of hyphae and breaks the links between plants and fungi. Deep ploughing disrupts the whole soil structure, bringing up subsoil and destroying humus-rich strata. While no-till practices protect the microorganisms in the soil when a crop is growing, harvesting and leaving the land fallow removes their source of nourishment. This is why sowing a cover crop after harvest is so important to continue the nutrient-rich conditions that the microorganisms need. Even fallowing of 2–3 weeks can damage mycorrhizae, as their hyphae die through lack of nourishment.

Fast-growing cover plants such as mustard can be used to replenish the soil until the slower-growing cover crops catch up. Using cover crops, especially legumes, between main crops helps maintain high levels of fungal spores and promotes good mycorrhizal development ready for the next crop. In gardens and vegetable plots, using good compost helps to maintain the soil microorganisms, and cover crops can be sown even in small gaps in cultivated beds. Healthy soil maintenance should always be a priority. Eliminating pesticides and minimizing digging in the garden is as much an advantage as no ploughing on the land – good for wildlife, people and the environment.

In general, there is no need to add extra, manufactured mycorrhizae, which may only contain a few species. These may also have a short shelf-life, as heat can kill active fungi. Looking after the soil and adding rotted manure and compost will maintain a healthy balance of mycorrhizae, and the rewards will be healthy crops.

The most common of the mycorrhizae are divided into the following.

Ectomycorrhizae

These form associations with roots in the rhizosphere (the narrow section of soil around the roots), without entering the cells. They are mostly found in the soil around different types of trees, such as pine, fir, spruce and oak. Thick hyphal strands called rhizomorphs can form, which are capable of conducting water, nutrients and chemical messages over long distances. Many of the fungi common in woodland are ectomycorrhizal fungi, with mushroom or toadstool-like fruiting bodies.

Endomycorrhizae

These enter the cells of plant roots. The fungi form structures within the cortical cells and also grow between the cells. The membranes of the fungus and the plant are in direct contact with each other. There are several types of endomycorrhizae, including arbuscular, ericoid and orchid mycorrhizae.

Arbuscular mycorrhizae

Arbuscular refers to the finger-like hyphae. They are some of the most ancient and widespread fungi in nature, efficiently exchanging nutrients with the plant hosts. These plants may show good tolerance to various stresses, as a result of the fungal activity. Arbuscular mycorrhizal fungi are the most important of the beneficial microorganisms in cultivated soils. Significant increases in the yield of crop plants have been noted following inoculation with arbuscular mycorrhizae.

Ericoid mycorrhizae

Species of the plant family Ericaceae tend to grow in sandy, acidic or even polluted ground, often as pioneer plants. The extra nutrients they need are provided by associating with ericoid myccorhizae, which have evolved to specialize in this family.

Orchid mycorrhiza

Orchids are totally dependent on fungi associated with their roots in order to grow. All orchids depend on the sugars provided by their fungal partner for at least some part of their existence. The dust-like orchid seeds require the nutrients provided by fungal invasion in order to germinate. Once the seedcoat ruptures and roots begin to emerge, the hyphae of orchidaceous mycorrhizae penetrate the root cells, creating coils, or pelotons, enabling nutrient exchange. The ghost orchid (*Epipogium aphyllum*) is a rare saprophyte, often existing for years underground, living on decaying matter and flowering unpredictably.

Saprophytes

Saprophytes are the decomposers, essential organisms that consume debris, corpses, faeces and general detritus. Without them, we would drown in a sea of excrement and dead bodies. These fungi convert manure into compost, which can then be put on the land. They cause decay, converting dead organic material into fungal biomass, carbon dioxide (CO_2), and small molecules, such as organic acids. These fungi can invade dead or damaged wood, and decompose cellulose and lignin. A few fungi are called 'sugar fungi' because they use the same simple substrates as do many bacteria. Like bacteria, fungi are important for fixing or retaining nutrients in the soil. In addition, many of the secondary metabolites of fungi are organic acids, so they help increase the rich organic matter that is resistant to degradation and may stay in the soil for hundreds of years. These fungi have enzymes that can 'rot' or 'digest' the cellulose and lignin found in organic matter. Lignin is an important source of carbon for many organisms. Without their digestive activities, organic material would continue to accumulate forever. These fungi have a widespread underground network of hyphae called mycelia, which bunch and branch together into fruiting bodies. These are the toadstools and mushrooms that appear in the autumn.

Pathogens

Pathogens cause infections, resulting in sickness or death when they invade plants, causing diseases such as potato and tomato blight, onion rot, rusts and many others. Pathogens can remain in the soil for long periods, waiting for the right plant.

Environmental conditions are important, as some fungi attack when warmth and humidity occur together. They can be adaptable, affecting different species of plants. Root rot is most commonly caused by species of *Cylindrocladium*, *Pythium*, *Phytophthora* and *Rhizoctonia*, which invade the root, blocking the flow of water and nutrients to the plant. The root system will eventually decay. *Fusarium* and *Verticillium* spp. cause wilting. The plant does not respond to watering. The effect can be dramatic, as in *Clematis* wilt. The stems can be fine and turgid one day and almost dead the next. Not much can be done except to trim off the dying stems and cover deeply with compost. The plant sometimes puts up healthy new stems. Damping off, which kills young seedlings, is caused by several fungi, including *Phytophthora* spp. Using clean tap water is an effective way of reducing damping off, which may spread with rainwater. Creeping thistle (*Cirsium arvense*) is attacked by the rust fungus *Puccinia obtegens*, which weakens and stunts its growth. The fungus occurs worldwide and has been evaluated in North America as a potential biocontrol agent. Infection may cause a local or a systemic problem. If a systemic infection becomes established in the rhizome system, the plant produces pale, stunted shoots that rarely flower. The fungus has caused epidemics that have eliminated creeping thistle from local areas. Resistance to the fungus has been found in some areas of Canada. The rust fungus *Puccinia punctiformis* may cause premature death of creeping thistle or at least prevent it flowering. In Germany, inoculation with the fungus in June combined with the effect of a cutting treatment reduced the number of fertile flower heads that developed. However, it has so far proved an unreliable way of eliminating this weed.

Actinomycetes

Actinomycetes are a large group of bacteria that grow as hyphae like fungi. They are responsible for the characteristically 'earthy' smell of freshly turned, healthy soil. Actinomycetes decompose a wide range of substrates, but are especially important in degrading recalcitrant (hard-to-decompose) compounds, such as chitin and cellulose, and are active at high soil pH levels. Fungi are more important in degrading these compounds at low soil pH. A number of antibiotics are produced by *Actinomycetes* such as *Streptomyces* spp.

Conclusion

There is a great deal of research into the actions of fungi, and in particular mycorrhizae, especially where soil health is concerned. Improving soil in a natural way can improve yield by up to 50%, thus reducing the reliance on artificial fertilizers. This also means research into more harmful pathogens, which can produce unpleasant diseases. A satisfactory result of the journey away from pesticides is the move towards regenerative farming, with its holistic approach to land and the environment. Watch this space.

Bibliography

Agricultural Marketing Resource Center (2018) Buckwheat. Available at: https://www.agmrc.org/commodities-products/grains-oilseeds/buckwheat (accessed 22 September 2021).

Bagyaraj, D.J. and Revanna, A. (2017) Soil biodiversity: role in sustainable horticulture. In: Peter, K.V. (ed.) *Biodiversity in Horticultural Crops*, Vol. 5. Daya Publishing House, New Delhi.

Björkman, T., Bellinder, R., Hahn, R. and Shail, J.W. Jr (2008) *Buckwheat Cover Crop Handbook*. Cornell University, Geneva, NY.

Garden Myths (2015) Mycorrhizae fungi inoculant products. *Garden Myths*, online. Available at: www.gardenmyths.com/mycorrhizae-fungi-inoculant-products (accessed 28 September 2021).

Lanfranco, L., Fiorilli, V. and Gutjahr, C. (2018) Partner communication and role of nutrients in the arbuscular mycorrhizal symbiosis. *New Phytologist* 220, 1031–1046.

Lewis, J.D. (2016) Evolution and diversification of mycorrhizal fungi. In: Kliman, R.M. (ed.) *Encyclopedia of Evolutionary Biology*. Elsevier, Oxford, pp. 94–99.

Magdoff, F. and van Es, H. (2021) *Building Soils for Better Crops: Ecological Management for Healthy Soils*. Sustainable Agricultural Research and Education, College Park, MD. Available at: www.sare.org/resources/building-soils-for-better-crops/ (accessed 11 October 2021).

Müller, E. and Nentwig, W. (2011) Plant pathogens as biocontrol agents of *Cirsium arvense* – an overestimated approach? *NeoBiota* 11, 1–24.

Royal Horticultural Society (2020) *Soil Care*. Royal Horticultural Society, London. Available at: www.rhs.org.uk/advice/plant-care/soil-care (accessed 11 October 2021).

Tedersoo, L., Bahram, M. and Zobel, M. (2020) How mycorrhizal associations drive plant populations and community biology. *Science* 367: eaba1223.

Thilagar, G. and Bagyaraj, D.J. (2015) Influence of different arbuscular mycorrhizal fungi on growth and yield of chilly. *Proceedings of the National Academy of Sciences, India Section B: Biological Sciences* 85, 71–75.

Zinkand, D. (ed.) (2010) Soil fungi are the root of all yields. *No-till Farmer*, online. Available at: http://cdn.shopify.com/s/files/1/0248/9641/files/FungiYields.pdf?1284 (accessed 28 September 2021).

6 Phytoremediators

Man is the only animal that fouls its own nest.

Anon.

Many heavy metals, such as mercury, cadmium, lead, nickel and chromium, are known to be toxic or carcinogenic. Public awareness of the damaging pollution caused by these metals is growing, but the problem of how to deal with the environmental damage is challenging, especially where the food chain is involved. Soil fertility is adversely affected, and conventional clean-up methods, either physical or chemical, are expensive and may create more problems for the environment. Trace metals are essential for healthy plant growth but become poisonous at high concentrations. Plant species that occur naturally in metalliferous soils have evolved mechanisms to tolerate various metals and to accumulate them in their tissues. There are different ways in which plants cope with toxins. Some block them, while others tolerate and absorb them into their tissues. Those species that are able to absorb heavy metals and other contaminants, thereby reducing the amount of toxic matter in the soil, can be suitable as phytoremediators.

The process of cleaning up polluted soil or water using plant stems and roots to extract contaminants is called phytoremediation. It is relatively inexpensive, eco-friendly and can prevent harmful substances from leaching into groundwater. It can be done without national or corporate grants, and does not need heavy equipment, which can further damage the environment. Washing soil with selected chemicals only causes more pollution, and a worst-case scenario is the complete removal of contaminated layers to landfill. Leaching out of these contaminants into watercourses causes environmental disruption and often widespread poisoning of fish and other creatures. Chemical systems that treat the land usually generate secondary waste and leave behind soil that has lost its fertility. Recovery can take many years.

Certain plants are able to trap and store heavy-metal toxins in their cells or contain them in their roots. It has been suggested that using edible plants that accumulate zinc can aid people with zinc deficiency. Phytate (inositol hexaphosphate salts) is the principal storage form of phosphorus in cereals, legumes and oleaginous seeds. It is abundant in the plant-based diets of many populations due to its presence in these staple foods. However, too much phytate in the diet can reduce the absorption of trace elements such as zinc and iron.

Mercury, which is incredibly toxic, can be absorbed by maize (*Zea mays*) and canola (*Brassica napus*). A number of plant species have been categorized as mercury hyperaccumulators, including the fern *Cyrtomium macrophyllum*, horsetail (*Equisetum telmateia*), red fescue (*Festuca rubra*) and hawkbit (*Leontodon taraxacoides*). Phytoremediation of mercury-contaminated sites has low environmental impact but will obviously take a long time. After harvest, the plant biomass containing mercury needs to be dealt with, and may cause health risks if not properly treated.

Sulphur lead, silver, gold, chromium, uranium, caesium and strontium, as well as damaging pesticides, can be cleaned up by suitable plants. It may be slow, but it is relatively cheap and an environmentally sound and sustainable method. Minute particles of gold, much in demand for recovery from electronic waste, can be retrieved by *Aspergillus niger*, a fungus found in compost, soil and other detritus.

DOI: 10.1079/9781789243604.0006

Aspen (*Populus* spp.) and mustard (*Brassica nigra*) can remove petroleum spills and lead from polluted ground. Sunflowers (*Helianthus annuus*) are able to remove radioactive materials from soil and water. Mercury, lead, cadmium, nickel and arsenic are among the hazardous cocktails that plants may need to absorb and neutralize. Toxin uptake is restricted to the reach of the plant roots, and there has to be some isolation of the polluted area to prevent stock or wildlife from grazing. While it is desirable to avoid causing contamination in the first place, once the damage has been done, phytoremediation can provide an environmentally sound method of cleaning shallow soil. Soils may be polluted with hydrocarbons, chlorinated solvents, pesticides, heavy metals and explosives and by leaching of agricultural products and landfills. Industries manufacturing paper and hardboard discharge various metals, mainly iron, manganese, zinc, copper, chromium, cadmium, nickel and lead, along with their waste products. Suitable plants can extract, degrade and immobilize these pollutants, and sometimes retrieve them for re-use (known as phytomining). Where the plants are burnt after harvesting, it is possible to obtain metal residue from the ash. In some cases, the best option may be to use plant roots to contain contaminated soil. The extensive root system filters the contaminants, drawing them up into their tissues. The plant biomass is regularly harvested, dried or incinerated.

There are several mechanical ways of dealing with pollution, such as incineration or thermal desorption (to remove organic or volatile contaminants from soil), particle size separation (to remove the soil fraction that holds the greater portion of the contaminants) and soil washing with selected chemicals (for the removal of organic or inorganic contaminants, by-products of laboratory processes). As a last resort, the complete removal of the contaminated material is packaged and sent to a landfill. These technologies lead to other difficulties, however. Excavation of contaminated material and the need to dispose of it in suitable landfill sites causes environmental disruption. Systems that treat the soil usually generate secondary waste (e.g. wash solution) and produce soil that has lost its fertility along with its contaminants. Not only are these methods very expensive, but the cleaning chemicals also degrade, producing further risks to the environment and animal health.

Disposal of Heavy Metals after Phytoaccumulation

After harvesting, plant biomass must be processed for removal of metals. Incineration of the biomass is probably the most effective means of disposal, reducing metals to ash, which can be treated as ores. At present, this may be uneconomical in small amounts. However, incineration can be used to generate energy, thus providing a useful commodity. The ash will still need to be disposed of carefully, but the original pollution should be much reduced.

Phytoextraction

Phytoextraction is the term for specific clean-up methods. It involves the absorption of contaminants by plant roots, which accumulate them up into stems and leaves. Plants called hyperaccumulators display a natural ability to tolerate high metal concentrations within their tissues. They are able to grow in heavily contaminated soils without displaying any or many toxic effects. Hyperaccumulators often contain high concentrations of several organic acids (e.g. citrate, malate and oxalate), which enable them to chelate the pollutants. The roots exude many compounds into the rhizosphere, including sugars, amino acids, phenolics and organic acids. Chelation means to grab or bind, and this is what happens when metals are accumulated in the plant tissues.

Petroleum Pollutants

Petroleum hydrocarbons that are in contact with the soil form a coating on the surface, preventing air and water circulation and causing suffocation of plant roots. The most effective industrial method of cleaning is soil washing, which involves scraping up the contaminated soil. The contaminants still need to be disposed of. Contaminated land from leaking pipelines in the Amazon is slowly being treated by local people using indigenous plants, fungi and bacteria. Aspen (*Populus* spp.) and willow (*Salix* spp.) are used successfully in Canada and other northern climates to clear land of oil pollutants.

Urban Environments

Urban soils are often damaged by a variety of materials and pollutants. These may be from small-scale industrial units, construction sites, and oil-related leaks and dumps (brown sites). There is no reason why contaminated urban soil should not be cleaned in a similar way to larger areas of damaged land. Population density may mean that gardens are not widely available. However, waste land, grass verges and rooftop spaces can be suitable for cultivation of fruit and vegetables. During difficult times, when food distribution is disrupted, it is even more important not to waste these spaces. It is easier to treat small areas of land with compost, cover crops, lawn clippings and leaf mould without too much expense. Phytoremediators come into their own in removing heavy metals and other unpleasant pollutants. The restoration of soil structure to a usable and safe medium will enable local communities to produce their own food, creating resilient supply systems.

Halophytes and Salinity

A halophyte is a plant that can tolerate salty conditions, coming into contact with salt water by its roots or from salt spray. Mangrove swamps, saltmarshes, deserts, seashores and lagoons are inhospitable places in which only the most adaptable plants can grow.

Halophytes in saltmarshes are adapted to water levels that fluctuate with the tide. Plants such as seagrass (*Zostera marina*), samphire (*Salicornia* spp.), cord-grass (*Spartina* spp.) and sea lavender (*Limonium* spp.) are important plants of coastal regions, providing shelter and nourishment for wildlife. Salinization is a worldwide problem, affecting approximately 1 billion ha of farmland. In the Netherlands, thousands of hectares could be salinized and rendered useless by 2030 due to rising sea levels, land subsidence and dry summers. Worldwide attempts have been made to halt this threat to food production, for instance by crop improvement and desalinizing land. Such experiments in the Netherlands have found a greater tolerance for salinity in certain crops, such as potato, carrot, cabbage and onion.

Plastics

Minute particles of various plastics are found practically everywhere, not just in the land and sea, but also in our bodies. This implies that crops are taking up these microplastics from irrigation and fertilizers. Polymethyl methacrylate (PMMA) is commonly used in plastic microbeads, and is also found in cosmetics and toothpastes. The minute size of these microbeads means that they can escape filtration and enter watercourses and eventually the sea. Recent research has identified a bacterium, *Pseudomonas* spp., that can digest some plastic compounds, using them as a source of energy. *Pseudomonas* is known for its ability to withstand harsh conditions, such as high temperatures and acidic environments. It may be one answer to the disposal of nuisance plastics.

A Selection of Plants Commonly Used for Phytoremediation

Aspen (*Populus* spp.; Salicaceae)

Of the many species in the genus *Populus*, the better known are the European aspen (*P. tremula*) and the closely related North American aspen (*P. tremuloides*). The roots of these trees can reduce certain contaminants to safe levels over long time frames. Rather than harvesting contaminants, phytostabilization may be a more viable option. Phytostabilization involves locking pollutants in the soil through accumulation in the root zone. Once the roots are mature enough to spread out, this process can be effective. In the interim, plants such as reed canary grass (*Phalaris arundinacea*) can be planted to provide some immediate stability to the surface soils and act as protection while trees are becoming established. Aspen has been used successfully in stabilizing land, as its roots continue to grow and spread even when the original tree has been cut down. Vegetative growth results in clones of trees, called ramets. Its ability to absorb zinc and trace elements allows aspen to survive on the spoil heaps of mines and in reclamation and landscaping of damaged land. Polluted areas, such as lignite mines in Germany, opencast oil shale quarries in Estonia and places contaminated by slag heaps, heavy metals and smelter pollution, can benefit from this treatment. In North America, *P. tremuloides* has colonized large areas of polluted industrial sites, tolerating acidic and poor soils. Aspen 'suckers' (vertical shoots from the roots) enable the plant to spread rapidly across an area. Some aspen forests have become diverse habitats by colonizing contaminated land near mining and smelting sites. *P. tremula* and other species of *Populus* have been recorded growing on the spoil heaps of abandoned arsenic mines in the UK. Aspen has a symbiotic relationship with a group of mycorrhizal fungi (ectomycorrhizae), which helps to increase phosphorus uptake, essential for plant growth. Using aspen is a long-term solution over many years.

Berseem or Egyptian clover (*Trifolium alexandrinum*; Leguminosae)

Egyptian clover is suitable for phytoextraction of cadmium, lead, copper and zinc owing to its fast growth, resistance to pollution, high biomass and multiple harvests in a single growth.

Brassica (Brassicaceae)

Brown mustard (*Brassica juncea*) has a large biomass making it more suitable than other species for accumulating certain metals. Ethiopian mustard (*Brassica carinata*) is an efficient hyperaccumulator of heavy metals such as cadmium, copper, nickel, chromium, arsenic and zinc.

Common reed (*Phragmites australis*; Poaceae)

Commonly seen around lake edges or in brackish water, reed thrives in a permaculture clean-up system. Reed beds form an aquatic plant-based system in which the bacteria in its roots feed on the organic matter in the sewage. Reed beds are employed for tertiary water treatment, or for refining sewage effluent to achieve stringent water-quality requirements. Reed beds are often used to reduce effluent flow rates from caravan parks or holiday venues into sensitive watercourses. Reed beds are constructed in artificial shallow ponds, which are constantly flooded. The effluent enters at one end and flows through the reeds' bacteria-rich root zone before being discharged on the far side. *P. australis* is able to transfer oxygen from its leaves, down through its stem and rhizomes into the gravel around the root system, thus creating a high population of microorganisms. The effluent percolates through layers of sand and gravel in an enclosed bed, and pollutants are broken down, turning toxins into nitrates. Reed beds need regular maintenance, to prevent the buildup of pathogens detrimental to health.

Moso bamboo (*Phyllostachys pubescens*; Poaceae)

Experiments have shown that this plant has a high tolerance of lead when grown in contaminated conditions. This could be a useful hyperaccumulator as, like many bamboos, it grows vigorously in large swathes.

Other species of bamboos are also useful. In addition to stabilizing soils with their invasive root system, bamboos contribute organic matter, and increase fertility, microbial biomass and the carbon content of soils. Because bamboo can thrive in problem soils and grows so fast, it is particularly efficient in repairing degraded land. Bamboo is also effective at absorbing and binding toxins such as heavy metals. These toxins are concentrated in the bamboo plant tissue, which must then be disposed of safely (as discussed earlier). Large bamboo stems are strong enough to be used as scaffolding and to build dwellings. Bamboo is also efficient at absorbing water polluted by effluent from farms or factories.

Pennycress (*Thlaspi caerulescens*; Brassicaceae)

This species grows in regions where metals occur naturally in the substrate. Over time, it has evolved the ability to tolerate and accumulate toxic levels of heavy metals. *T. caerulescens* has been studied extensively for its ability to hyperaccumulate zinc, cadmium and nickel. As much as 30,000 ppm of zinc and 10,000 ppm of cadmium in the shoot biomass have been recorded, without any signs of toxicity. However, its small habit makes it a long-term project as a hyperaccumulator, as it may take many years to clean up polluted land.

Reed canary grass (*Phalaris arundinacea*; Poaceae)

P. arundinacea is a tall, reed-like grass growing up to 200 cm high, with sturdy, smooth stems. Its creeping rhizomes enable it to form dense stands. Its panicles can be up to 25 cm long, with spikelets sometimes flushed purple. It grows in damp or wet places, and can accumulate lead and zinc in its tissues.

Sunflower (*Helianthus annuus*; Asteraceae)

Sunflower is a tall, sun-loving, beautiful annual, with a capitulum composed of many florets, both disc (inner florets) and ray (outer petaloid florets). It gets its name from the way the flower head turns towards the sun, following it on its circuit. It is a useful phytoremediator and hyperaccumulator, with a strong wide-spreading root system. As a fast growing annual, it can be planted in large stands during summer and then harvested with its collection of contaminants. Adding quantities of composted manure when sowing greatly improves the plant's ability to grow and absorb toxins. Famously planted in the extremely contaminated land around Chernobyl, these plants continued to thrive in the nuclear desolation. Not only were heavy metals absorbed, but the plants also ingested toxins in water from contaminated ponds in the area. In Japan, the Serakogen Farm in Hiroshima grows over a million sunflowers. This part of the devastated territory is now safe to visit and has become a tourist attraction. People flock to the site in summer to admire the sunflowers in bloom, and to remember those who perished there.

Tobacco (*Nicotiana tabacum*; Solanaceae)

This species is an important economic plant, with leaves containing high quantities of nicotine, suitable for the manufacture of cigarettes and cigars. It is a tall, bulky and fast growing plant, and its poisonous nicotine content deters herbivores and other pests. After the removal of the attractive flowers, known as 'topping', nicotine becomes even more concentrated in the leaves, and is ready for harvesting and drying.

Because of widespread health problems caused by smoking tobacco, demand for the product has dropped. Scientists are looking for alternative uses for it, including for phytoremediation and as a natural pesticide. For many years, gardeners have used home-made mixtures of tobacco and water as a way of killing insect pests. A 'green' pesticide industry based on tobacco could provide additional income for farmers, as well as an environmentally friendly pest control. The nicotine is produced by the plant when it is stressed by predator attack and is thought to give it some measure of protection. In polluted land, *N. tabacum* displays high tolerance for heavy metals and, because of its rapid growth and high biomass, is regarded as a useful phytoremediator. Its roots have been shown to accumulate cobalt, nickel, copper and cadmium, and the leaves can assimilate zinc, selenium and mercury. While the tobacco industry is not seeking to use contaminant-laden crops in its products, its research into varieties that absorb fewer substances may uncover interesting and useful information.

Willow (*Salix* spp.; Salicaceae)

The power of willow trees to purify polluted water has been known for some time., A dense planting of willow is ecologically sound and can be used to clean foul water and remove a high percentage of nitrates and phosphates. It can be planted around sewage-treatment works to purify treated water and clean activated sludge. Water is purified when it passes through the roots of densely planted willow trees. Willow grows well in this moist habitat. A useful by-product of its strong growth is the harvesting of trees after several years for burning in power stations or biomass boilers. Willow branches are flexible and are popular for basket making (osier willow (*S. viminalis*)). Willow wood (*S. alba* var. *caerulea*) is famous for making cricket bats.

Bibliography

Ali, H., Naseer, M. and Sajad M.A. (2012) Phytoremediation of heavy metals by *Trifolium alexandrinum*. *International Journal of Environmental Science* 2, 1459–1469.

Argumedo-Delira, R., Gomez-Martinez, M.J. and Soto, B.J. (2019) Gold bioleaching from printed circuit boards of mobile phones by *Aspergillus niger* in a culture without agitation and with glucose as a carbon source. *Metals* 9: 521.

Assunção, A.G.L., Schat, H. and Aarts, M.G.M. (2003) *Thlaspi caerulescens*, an attractive model species to study heavy metal hyperaccumulation in plants. *New Phytologist* 159, 351–360.

Cárdenas Espinosa, M.J., Colina Blanco, A., Schmidgall, T., Atanasoff-Kardjalieff, A.K., Kappelmeyer, U. *et al.* (2020) Toward biorecycling: isolation of a soil bacterium that grows on a polyurethane oligomer and monomer. *Frontiers in Microbiology* 11: 404.

Centre for Alternative Technology (2020) How is a reed bed sized and designed. Centre for Alternative Technology, Powys, UK. Available at: https://cat.org.uk/info-resources/free-information-service/water-and-sanitation/sewage-treatment/ (accessed 28 September 2021).

Cheeseman, J.M. (2015) The evolution of halophytes, glycophytes and crops, and its implications for food security under saline conditions. *New Phytologist* 206, 557–570.

Gajić, G., Djurdjević, L., Kostić, O., Jarić, S., Mitrović, M. and Pavlović, P. (2018) Ecological Potential of plants for phytoremediation and ecorestoration of fly ash deposits and mine waste. *Frontiers in Environmental Science* 6: 124.

Lal, R. (2020) Home gardening and urban agriculture for advancing food and nutritional security in response to the COVID-19 pandemic. *Food Security* 12, 871–876.

Liu, D., Li, S., Islam, E., Chen, J., Wu, J. *et al.* (2015) Lead accumulation and tolerance of Moso bamboo (*Phyllostachys pubescens*) seedlings: applications of phytoremediation. *Journal of Zhejiang University – Science B: Biomedicine & Biotechnology* 16, 123–130.

Liu, H., Wang, H., Zhang, Y., Wang, H., Yang, J. *et al.* (2019) Comparison of heavy metal accumulation and cadmium phytoextraction rates among ten leading tobacco (*Nicotiana tabacum* L.) cultivars in China. *International Journal of Phytoremediation* 21, 699–706.

Mackenzie, N. (2010) *Ecology, Conservation and Management of Aspen.* Scottish Native Woods, Aberfeldy, UK.

Mahardika, G., Rinanti, A. and Fachrul, M.F. (2018) Phytoremediation of heavy metal copper (Cu^{2+}) by sunflower *(Helianthus annuus L.). Earth and Environmental Science* 106: 012120.

Salt Farm Foundation (2017) Breakthrough in food security with Dutch knowledge of saline agriculture. Salt Farm Foundation, Den Hoorn, Netherlands. Available at: https://saltfarmfoundation.com/external-news/breakthrough-in-food-security-with-dutch-knowledge-of-saline-agriculture/ (accessed 28 September 2021).

Wang, L., Hou, D., Cao, Y., Ok, Y.S., Tack, F.M.G. *et al.* (2020) Remediation of mercury contaminated soil, water and air. *Environmental International* 134: 105281.

Wikiwater (2012) A25 - Treatment by willow-plantation or reed-bed zones at the outlet of a sewage treatment plant. Available at: https://wikiwater.fr/a25-treatment-by-willow-plantation (accessed 28 September 2021).

Yazdanbakhsh, A., Alavi, S.N., Valadabadi, S.A., Karimi, F. and Karimi, Z. (2020) Heavy metals uptake of salty soils by ornamental sunflower using cow manure and biosolids: a case study in Alborz city, Iran. *Air, Soil and Water Research* 13: 117862211989846.

7 Illustrations, with Information on Each Plant

DOI: 10.1079/9781789243604.0007

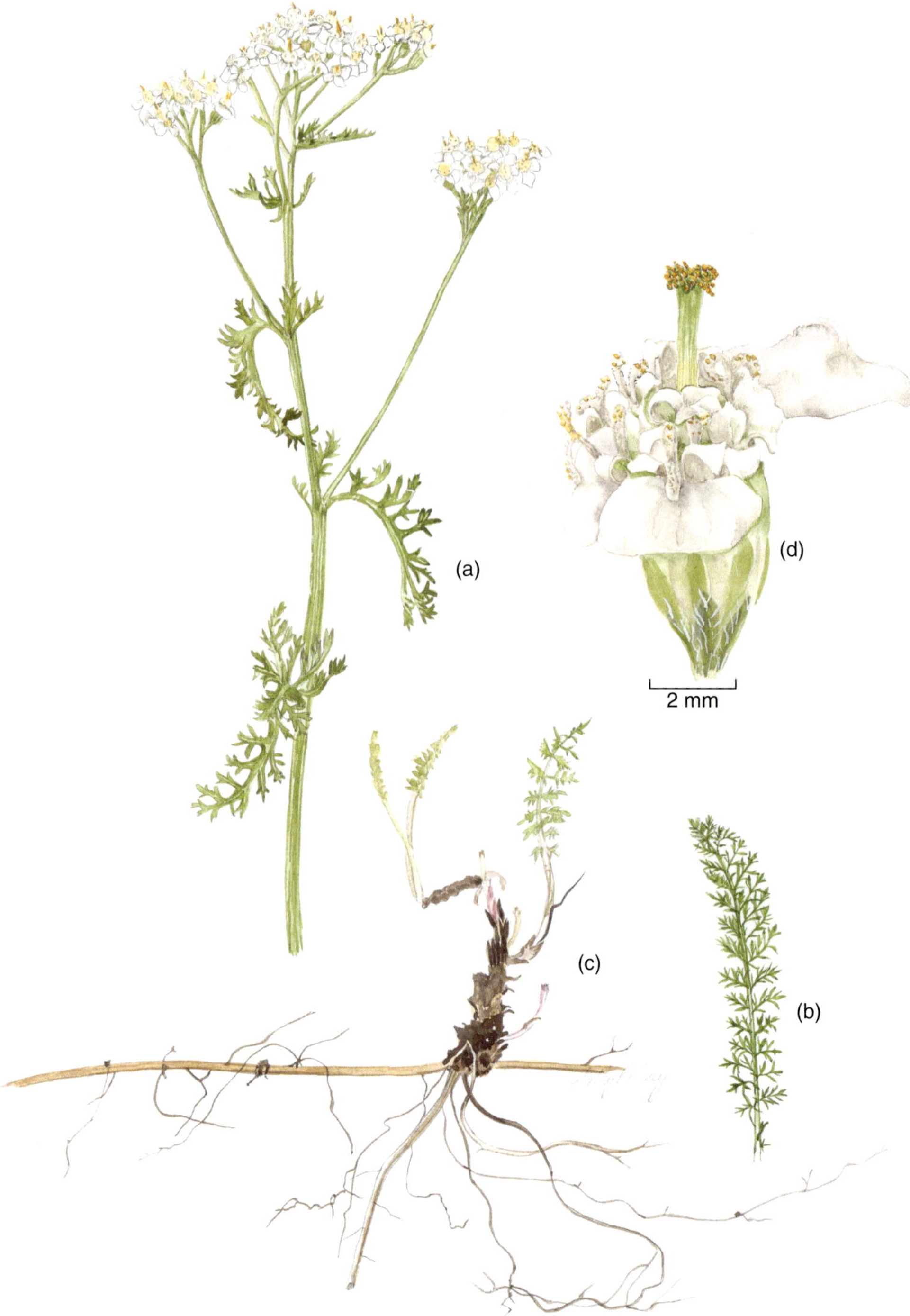

Fig. 7.1. *Achillea millefolium* habit (a), seedling leaf (b), root (c) and flower enlargement (d).

Achillea millefolium (Fig. 7.1)

Common name: Yarrow.
Main uses: Forage herb; deep rooting, persistent plant.

Environmental benefits

Growing, climate, temperature and rainfall requirements: Temperate climate and up to 1100 m altitude; drought tolerant.
Soil requirements: Suits all soils except waterlogged or acid soils.
Soil condition: Develops strong, deep roots, which improve drainage and break up soil compactions.
Insects: Food source for beetles, butterflies, hover flies and many other insects.
Grazing animals: High in protein and contains the volatile oils cineole, eugenol, camphor and azulene as well as the tannins salicylic acid and isovaleric acid. Acts as an antiseptic and defence against internal worms.

Structure

Roots: Deep, branched tap roots and smaller laterals.
Stems: Grooved and slightly woody; up to 65 cm in height.
Leaves: Deeply incised, feathery; arranged in a spiral.
Flowers: Flat clusters of white flowers; some may be pink.
Seeds: Rounded oblong.

Fig. 7.2. *Alnus glutinosa* leaf and immature cone (a), winter cone (b), catkin (c) and young bark (d).

Alnus glutinosa (Fig. 7.2)

Common name: Alder.
Main uses: Stabilizing watercourse margins, flood control, water purification, windbreaks, hedging, furniture, charcoal and pile timber.
Unsuitability: Susceptible to *Phytophthora alni*, a pathogen causing yellow leaves and trunk spots near the base, sometimes killing the tree. It is a robust colonizer classed as an undesirable in New Zealand where it can outcompete native species.

Environmental benefits

Growing, climate, temperature and rainfall requirements: Native to Europe, North Africa and South-west Asia.
Soil requirements: Tolerant of poor and waterlogged soils, but can thrive in drier places once established.
Soil conditioning: Hosts nitrogen-fixing bacteria and has associations with up to 47 species of mycorrhiza, so improves soil fertility beyond the addition of leaf litter. When planted in a mixed fruit and nut grove, it benefits the other trees.
Insects: A food plant of the larvae of many butterflies and moths and some 140 species of plant-eating insects.
Grazing animals: Grazed by cows, sheep and goats, and has medicinal qualities in part due to the high tannin content. Pigs do not like it and it is thought to harm horses.

Structure

Roots: Have a system of air ducts that supply oxygen internally for survival in water.
Stems: The lower bark contains cork cells with additional breathing capacity.
Leaves: Dark green and leathery, often dropping while still green, adding to the richness of leaf litter.
Flowers: Monoecious, with dark red male catkins and upright green female cones.
Seeds: Small and spread by water. Eaten by birds in winter.

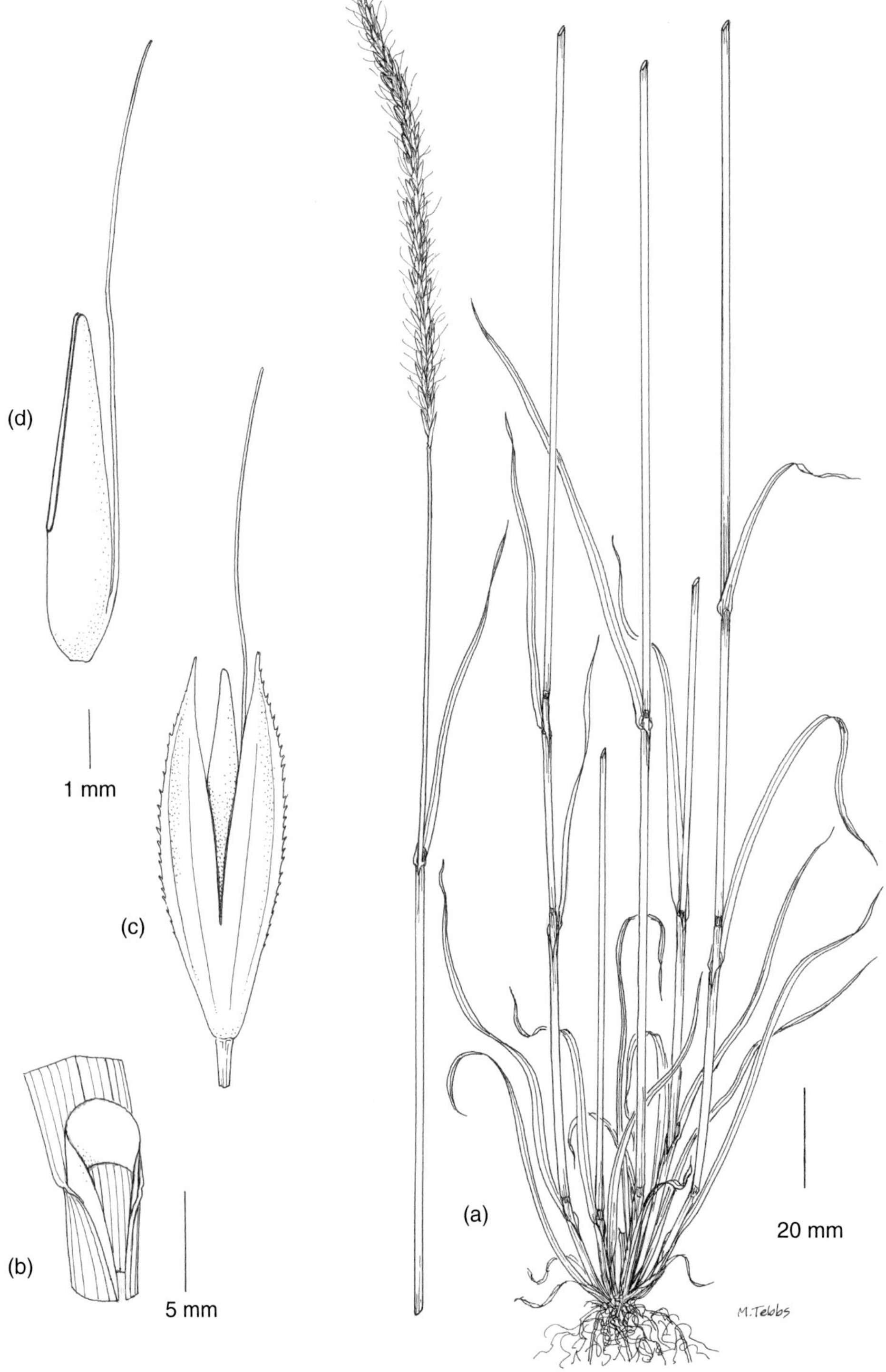

Fig. 7.3. *Alopecurus myosuroides* habit (a), ligule (b), spikelet (c) and floret (d).

Alopecurus myosuroides (Fig. 7.3)

Common name: Black-grass.
Unsuitability: Pernicious weed of arable land, thriving on damp, heavy soils, and germinating at the same time as winter wheats; very common.

Environmental benefits

Growing, climate, temperature and rainfall requirements: Annual frost-tolerant plant of open habitats. A troublesome weed among cereal crops.
Soil requirements: Thrives in moist, heavy, ill-drained soil. It has become tolerant to some herbicides and is difficult to get rid of with conventional methods.
Soil conditioning: Not welcome in agricultural soils.
Grazing animals: Tends to be avoided.

Structure

Roots: Fibrous.
Stems: Up to 90 cm tall.
Leaves: Blades flat, smooth and glabrous.
Flowers: Cylindrical panicle up to 13 cm long, tapering at the tip.
Seeds: Viable for at least 4 years.

Fig. 7.4. *Ammophila arenaria* habit (a) and flower (b); *Leymus arenarius* (lyme grass) habit (c), flower (d), leaf section (e) and root (f); *Elymus farctus* (sand couch) (g); *Carex arenaria* (sand sedge) (h).

Ammophila arenaria (Fig. 7.4)

Common name: Marram grass (illustration on frontispiece).
Main uses: Important in the stabilization of wind-blown sand.

Environmental benefits

Growing, climate, temperature and rainfall requirements: Mobile dune systems. Widely introduced to help stabilize loose sand.
Soil requirements: Marram grass is not very salt tolerant and will not take tidal flooding. It does best in dry dune soils that are low in nutrients and organic content.
Soil conditioning: Widely planted to stabilize and prevent dune erosion where its rapidly spreading rhizomes grow through the sand, binding it. Once the dunes are fixed, it tends to flower less, but remains vegetatively vigorous.
Grazing animals: Not grazed.

Structure

Roots: Deeply penetrating rhizomes.
Stems: Up to 90 cm tall.
Leaves: Stiff, greyish-green, tightly inrolled.
Flowers: Spike-like panicles.
Seeds: Many seeds are produced, but mortality is high.
Marram grass can hybridize with *Calamagrostis epigejos* (wood small-reed), producing a sterile hybrid (× *Calammophila baltica*).

Fig. 7.5. *Avena sativa*.

Avena sativa (Fig. 7.5)

Common name: Oat.
Main uses: Cover crop and green manure, and as a cultivated cereal crop.

Environmental benefits

Growing, climate, temperature and rainfall requirements: Annual, tolerating cold, damp conditions. Suitable for northern climates.
Soil requirements: Tolerates poor, moist soil.
Soil conditioning: When left on the surface, the oats provide cover and the dead stems help trap and conserve moisture. Useful as a weed suppressant.
Grazing animals: Used as fodder for horses.

Structure

Roots: Fibrous.
Stems: Up to 130 cm tall.
Leaves: Flat, rough blades.
Flowers: Nodding panicle.
Seeds: Caryopses.

Fig. 7.6. *Avicennia marina* flower (a), fruit (b), seed (c), seedling (d), leaf (e), tree habit sketch (f), pneumatophore (g), cross-section of pneumatophore (h), root (i) and cross-section of root (j).

Avicennia marina (Fig. 7.6)

Common name: Grey mangrove.
Main uses: Coastal protection, marine species protection, biodiversity enhancer.
Unsuitability: Only suitable for warmer climates.

Environmental benefits

Growing, climate, temperature and rainfall requirements: This species of mangrove will survive in drier climates and in baked salt pans.
Soil requirements: Grows in river estuaries, tidal areas and salt marshes.
Soil conditioning: Contributes rich leaf litter, which decomposes quickly and supports a dynamic ecosystem. Efficient nutrient recyclers and nitrogen fixers. Filters pollutants and consolidates sediments.
Insects: Bees may be one of the many pollinators. Ants protect the trees from other grazers.
Grazing animals: Not grazed.

Structure

Roots: Buttress roots support the tree in tidal variations. Pneumataphores supply oxygen in anaerobic mud.
Stems: Woody.
Leaves: Glossy and covered in a protective cuticle; the underside has pores that can exude excess salt.
Flowers: Creamy, white, four petals.
Seeds: The size of a broad bean. Seeds begin to germinate on the parent plant and are dispersed in the water, but only viable for a few days.

Fig. 7.7. *Calendula officinalis* flower (a), flower buds (b), seed head (c), seeds (d) and root (e).

Calendula officinalis (Fig. 7.7)

Common name: Marigold.
Main uses: Pot herb, cosmetic, medicinal and ornamental. Useful for attracting pollinating insects to vegetable plots.

Environmental benefits

Growing, climate, temperature and rainfall requirements: Likes full sun, fairly drought tolerant and hardy. Often suffers from mildew later in the season.
Soil requirements: Will grow on waste ground and rubbish tips, as well as in gardens.
Insects: Very attractive to insects.
Grazing animals: Not grazed.

Structure

Roots: Fibrous or with short tap root.
Stems: Up to 50 cm tall.
Leaves: Simple.
Flowers: Bright yellow capitulum, with ray and disc florets.
Seeds: Large, variable structures.

Fig. 7.8. *Cichorium intybus* flowering stem (a), habit sketch (b), seedling (c) and root (d).

Cichorium intybus (Fig. 7.8)

Common name: Chicory.
Main uses: Mixed ley with *Trifolium repens* (white clover) and *Lolium perenne* (perennial ryegrass), best for nutritious strip grazing. Has anthelmintic properties. Breaks up compacted soil to improve drainage.
Unsuitability: Perennial varieties can outcompete sainfoin; taller plants are less nutritious.

Environmental benefits

Growing, climate, temperature and rainfall requirements: Sow in spring or early autumn into warm soil. The grown plant is frost hardy.
Soil requirements: Should be rolled on sowing to increase water contact with the seed.
Soil conditioning: Improves drainage.
Insects: Good pollinator plant.
Grazing animals: Cattle and sheep are grazed; especially good for fattening lambs.

Structure

Roots: Strong and deep tap root.
Stems: Become woody with age if not grazed or cut.
Leaves: Broad in a basal rosette, with hairy, bluntly toothed margins.
Flowers: Flowers from mid-summer to early autumn.
Seeds: Small; a favourite with goldfinches.

Fig. 7.9. *Cirsium arvense* habit (a), flower (b), seedling (c), root (d), seed (e) and seed head (f); painted lady butterfly (g); painted lady chrysalis (h); painted lady caterpillar (i).

Cirsium arvense (Fig. 7.9)

Common name: Creeping thistle.
Main uses: Young leaves are edible to sheep, ponies, donkeys, goats and llamas. Contains copper and iron trace elements.
Unsuitability: Can spread and form infestations in arable crops.

Environmental effects

Growing, climate, temperature and rainfall requirements: Frequent in hedges and field margins. It can infest arable crops. It is frost hardy and prefers cooler climates without high summer temperatures. It thrives on aerated soils and will be absent where there is compaction and waterlogging. It is checked by drought.
Soil requirements: Tolerant of most soils and some salinity. Growth is stimulated by nitrogen fertilization.
Insects: Beneficial to pollinators. Eaten by flea beetles (*Altica carduorum*), leaf beetles (*Lema cyanella*), stem-mining weevil (*Hadroplontus litura*), thistle stem gall fly (*Urophora cardui*), painted lady butterfly (*Cynthia cardui*) and several leaf-feeding moth larvae. Host to the insect pests bean aphis, mangold fly, celery fly and swift moth larvae.
Grazing animals: Generally avoided by grazing animals if alternatives are available. It is harmless.

Structure

Roots: Long, thick vertical and horizontal roots, short fine shoots, adventitious roots.
Stems: Prickly and becoming woody with age.
Leaves: Deeply incised with spines. Basal rosettes may have differing shaped leaves.
Flowers: Purple, composite flower.
Seeds: Long, with pappus.

Fig. 7.10. *Corylus maxima* 'Purpurea' (purple filbert) catkin and female flower (a), nut cluster and leaf (b) and vertical section through nut (c).

Corylus avellana and *Corylus maxima* (Fig. 7.10)

Common names: Hazel nut/cob nut and filbert.
Main uses: Culinary crop, hedging.

Environmental benefits

Growing, climate, temperature and rainfall requirements: Native to the Middle East, it crops best in latitudes 50–35°, and once established is tolerant of dry summers and cold winters. It can be propagated from sucker cuttings and by layering.
Soil requirements: Tolerant of a wide range of soils. Can be grown on nitrogen-poor soils.
Soil conditioning: Hedges and ground cover prevent erosion and provide cover for wildlife.
Insects: The nut weevil (*Curculio nucum*) is an insect pest that leaves its larvae in the nuts. The catkins are an early provider of pollen for foraging bees.
Grazing animals: Not normally browsed.

Structure

Roots: Strong and deep, fibrous.
Stems: Wooden.
Leaves: Large, roundish, toothed.
Flowers: Monoecious; male catkins and female flowers on same branch.
Seeds: Large, nutritious, containing up to 17% protein and up to 15% fibre, plus vitamins B_1, B_6 and E, calcium and fat.

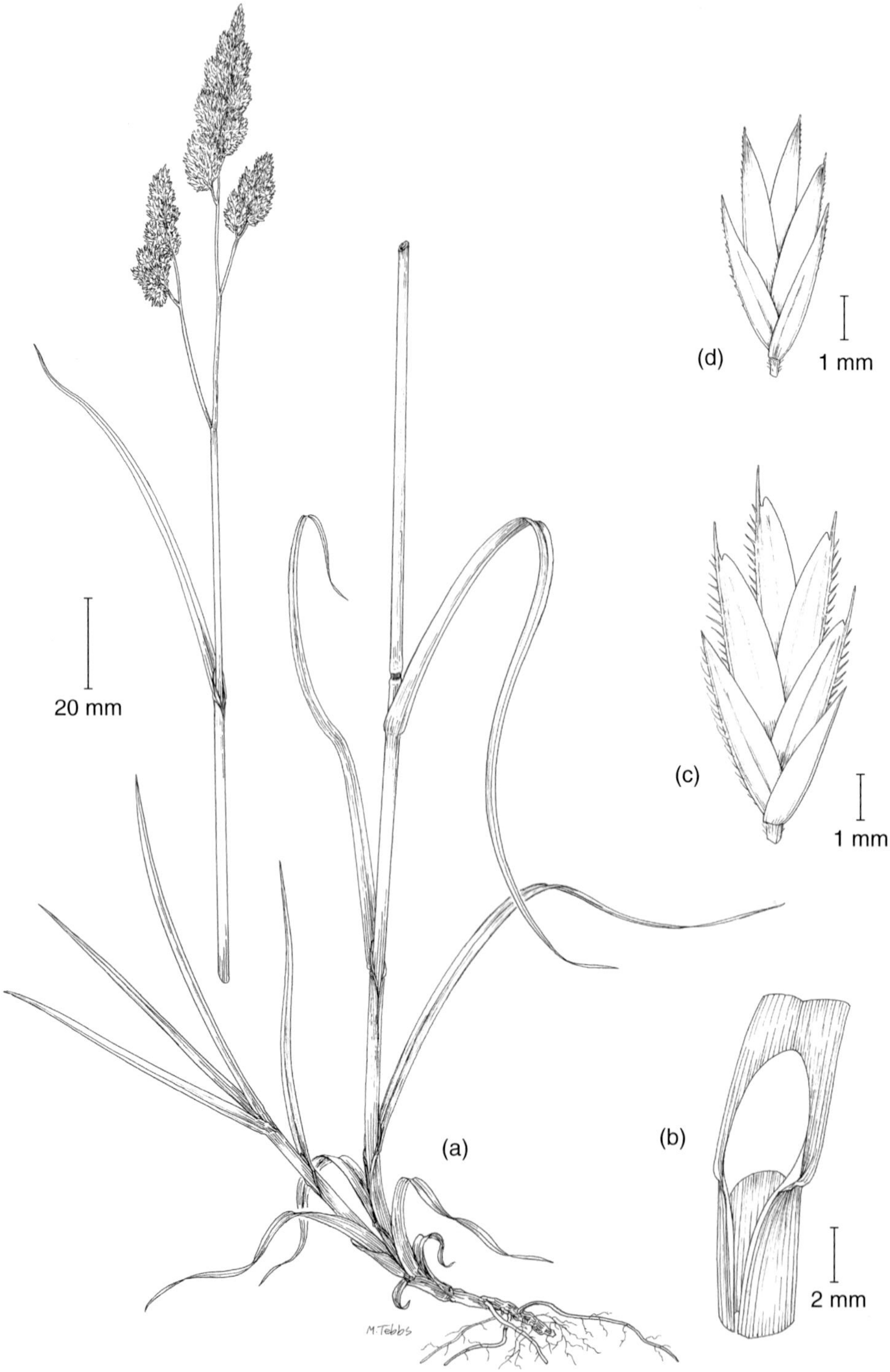

Fig. 7.11. *Dactylis glomerata* habit (a), ligule (b) and spikelet (c, d).

Dactylis glomerata (Fig. 7.11)

Common name: Cock's-foot.
Main uses: Cock's-foot has long been cultivated and selected for grazing.
Unsuitability: Unsuited for wet areas.

Environmental benefits

Growing, climate, temperature and rainfall requirements: Meadows, downland, woods, river banks, roadsides and waste ground. Can form large tussocks in unmanaged areas. Less common at higher altitudes.
Soil requirements: Neutral to basic soils, pH 5.0–8.0.
Soil conditioning: Meadow grass.
Grazing animals: Selected for grazing and hay.

Structure

Roots: Fibrous.
Stems: Coarse, densely tufted, with stems up to 120 cm tall.
Leaves: Mostly glabrous, smooth or slightly scabrid.
Flowers: Green, purple or yellowish panicle.
Seeds: May persist for 2–3 years.

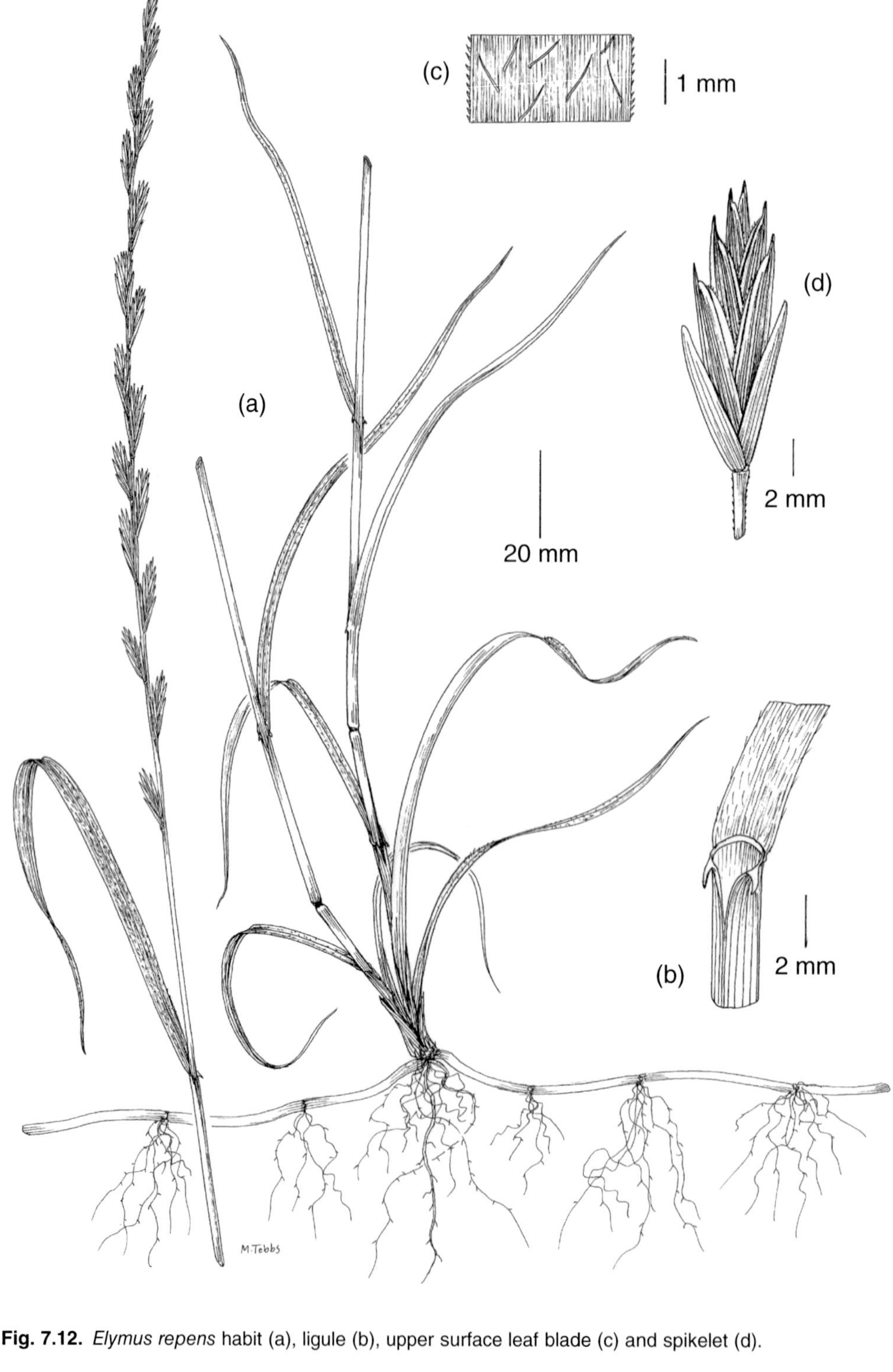

Fig. 7.12. *Elymus repens* habit (a), ligule (b), upper surface leaf blade (c) and spikelet (d).

Elymus repens (Fig. 7.12)

Common name: Common couch.
Unsuitability: Aoxious weed of gardens, orchards and arable land. A successful perennial, widespread and common in disturbed ground.

Environmental benefits

Growing, climate, temperature and rainfall requirements: Couch is at home in both light and heavy soils, but spreads more readily in tilled light soils that are low in organic content, but high in nitrogen. It is only a problem in cereal crops, peas and oilseed rape, where it reduces the crop yield. Grows in most habitats from coastal to hedgerows and waste ground.
Soil requirements: Disturbed soil; drought and cold tolerant.
Grazing animals: It is palatable and, when grazed, ceases to be a problem.

Structure

Roots: Long, stout, pointed rhizomes, which can regenerate from cut or broken pieces to form new plants.
Stems: Erect, slender.
Leaves: Flat to inrolled.
Flowers: Stiff, straight raceme.
Seeds: Seed production low; mostly reproduces vegetatively.

Fig. 7.13. *Equisetum arvense* habit (a), developing shoots (b) and cone-bearing fertile stem (c).

Equisetum arvense (Fig. 7.13)

Common name: Field horsetail
Main uses: Extracts are used as a fungicide. Accumulates the heavy metals, cadmium, copper, lead and zinc.
Unsuitability: Toxic to sheep cattle and horses. A problem weed on arable and grassland, orchards and nurseries. Hard to eradicate.

Environmental benefits

Growing, climate, temperature and rainfall requirements: Grows in temperatures from <5°C to >20°C and with annual rainfall between 100 and 2000 mm. Prefers moist conditions. Withstands flooding but not drought.
Soil requirements: Wide range of soil types, preferring stone-free hedge banks and waste land with low nutrient levels.
Soil conditioning: Deep roots may aerate waterlogged soils. Can decontaminate soils from heavy metals over time.
Grazing animals: Should not be grazed. Toxic both when fresh and particularly in hay. Horses are very sensitive to it.

Structure

Roots: Deep and lateral tubers.
Stems: Segmented and brittle; very rough with silica content.
Leaves: Segmented and narrow.
Flowers: Cone-bearing fertile stems.
Seeds: Spores.

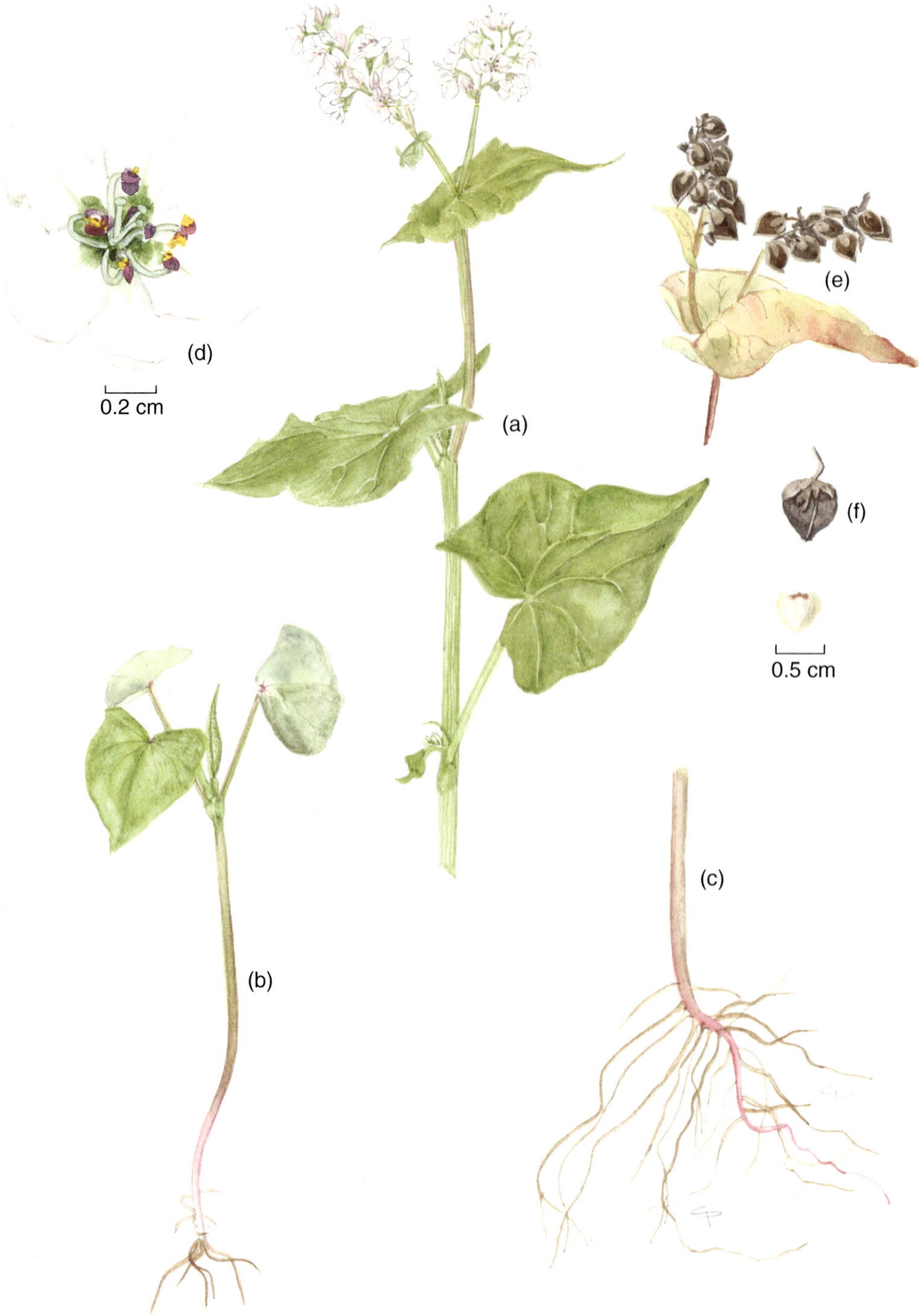

Fig. 7.14. *Fagopyrum esculentum* habit (a), seedling (b), root (c), flower enlargement (d), seed (e) and seed enlargement (f).

Fagopyrum esculentum (Fig. 7.14)

Common name: Buckwheat.
Main uses: Cereal with high nutritional qualities. Cover crop.
Unsuitability: Does not do well on soil where nitrogen fertilizers have been used. Not suitable for hot climates, unless sown late in the season.

Environmental benefits

Growing, climate, temperature and rainfall requirements: Suited to high latitude or northern areas. Grows 75–125 cm tall in 8–12 weeks. Not frost hardy.
Soil requirements: Does best on well-drained, low-fertility or acidic soils.
Soil conditioning: Restores soils with phosphorous lock-out. Suppresses weeds. Good biomass for green manure.
Insects: Good for pollinators. Hosts aphid predators.
Grazing animals: Not suited to pasture.

Structure

Roots: Main taproot with lateral shoots.
Stems: Reddish and fibrous as seed develops.
Leaves: Pale green cordate/sagittate in shape.
Flowers: White or pink.
Seeds: Triangular.

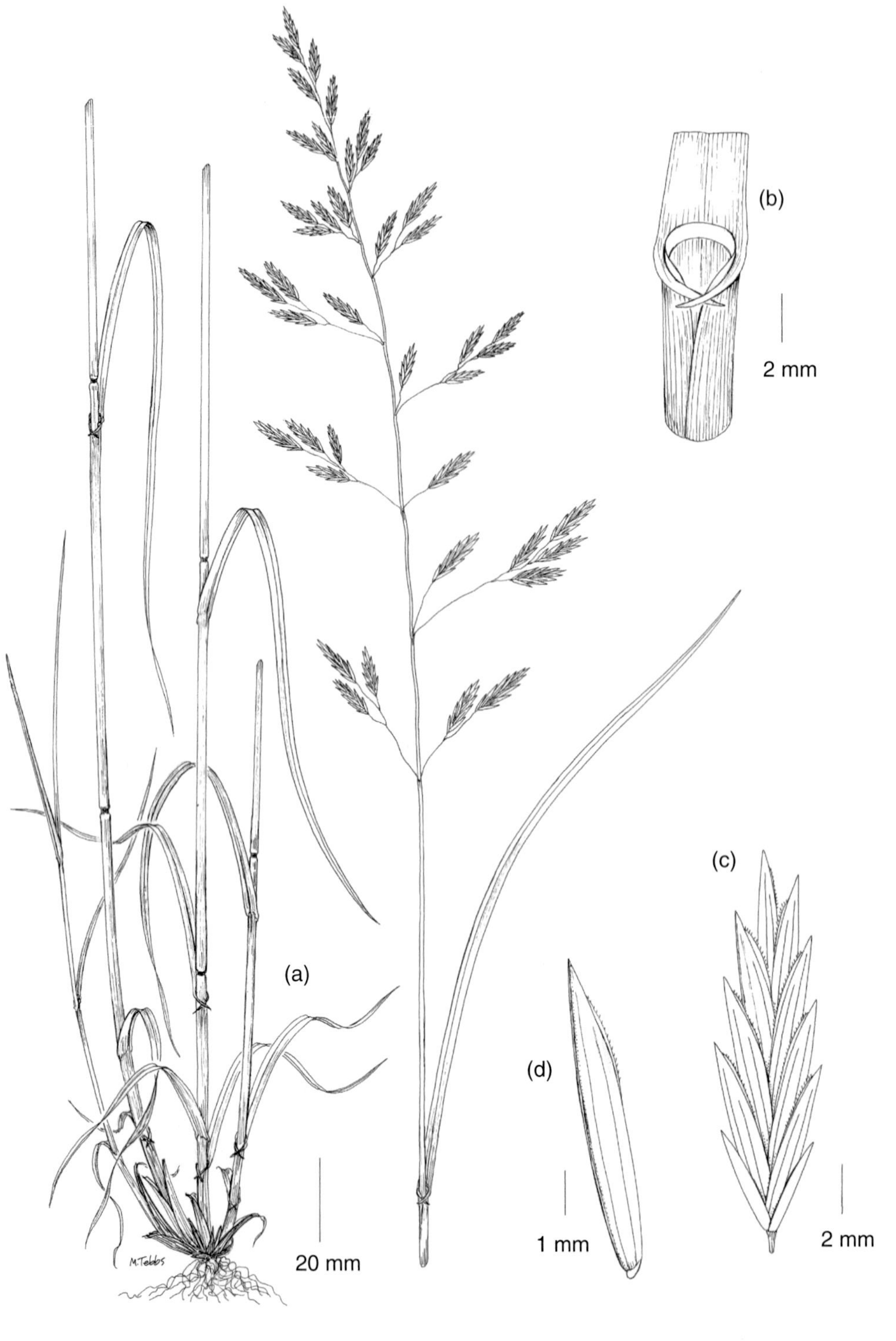

Fig. 7.15. *Festuca pratensis* habit (a), ligule (b), spikelet (c) and lemma (d).

Festuca pratensis (Fig. 7.15)

Common name: Meadow fescue.
Main uses: Loosely tufted perennial, growing in leys and water meadows with clovers and other grasses.
Unsuitability: Its popularity as an agricultural grass has declined in favour of *Festuca arundinacea*.

Environmental benefits

Growing, climate, temperature and rainfall requirements: Found in moist grassland habitats, including hay meadows and pasture. Grows from sea level to >500 m. Often found in species-rich areas. Highly cold tolerant. Useful for creating flower-rich grassland.
Soil requirements: Grows best in fertile or loamy moist soils above pH 5.0.
Soil conditioning: Grown as a companion plant with clovers and other grasses.
Grazing animals: Suitable for grazing or hay.

Structure

Roots: Fibrous roots.
Stems: Slender, unbranched, up to 100 cm tall.
Leaves: Blades flat, scabrous above, glossy and glabous below.
Flowers: Erect or nodding, loose panicle.
Seed: Caryopsis
Similar species: Hybridizes with *Lolium perenne* (× *Festulolium loliaceum*), and with *Lolium multiflorum* (× *Festulolium braunii*).

Fig. 7.16. *Helianthus annuus* 'Solar Flash' F_1 hybrid, hardy annual (a), 'American Giant' (b) and disc florets (c).

Helianthus annuus (Fig. 7.16)

Common name: Sunflower
Main uses: Seed crop; phytoremediator.
Unsuitability: Will not grow in a cold, dark or wet environment.

Environmental benefits

Growing, climate, temperature and rainfall requirements: Can be sown when temperatures reach 21°C. Likes full sun and some moisture but can tolerate low rainfall. Quick growing.
Soil requirements: A wide range of soil types, with a preference for pH 6–7.5.
Soil conditioning: Removes toxins and heavy metals from the soil, notably lead, arsenic and uranium. Used to clean up land after the Chernobyl and Fukushima nuclear disasters.
Insects: Insect pollinated. Good for bees and other pollinators. The seeds are enjoyed by birds.
Grazing animals: Not grazed, but birds eat the seeds.

Structure

Roots: Single, deep tap root with smaller laterals.
Stems: Rough, sturdy, hairy and ribbed; partly hollow for quick growth.
Leaves: Heart-shaped or oval, usually hairy.
Flowers: Central disc florets with ray florets on the perimeter.
Seeds: Oblong, sometimes striped (0.75–1.25 cm).

Fig. 7.17. *Hippophae rhamnoides*: fruiting branch (a), male flowers (b), female flowers (c), male flower enlargement (d) and female flower enlargement (e).

Hippophae rhamnoides (Fig. 7.17)

Common name: Sea buckthorn.
Main uses: Hedging and windbreaks, soil stabilization and nitrogen fixing. A superfood crop.
Unsuitability: Needs controlling as it can spread beyond planted areas. Tricky to harvest.

Environmental benefits

Growing, climate, temperature and rainfall requirements: Grows at low temperature and is damp tolerant. Bare-root plants should be planted during the winter in a proportion of one male plant to every ten female plants. It can also be grown from seed or cuttings but needs to be in the open.
Soil requirements: Tolerant of poor and sandy soils.
Soil conditioning: Fixes nitrogen and binds loose sandy soil.
Insects: Wind pollinated but a haven for wildlife.
Grazing animals: Not normally grazed.

Structure

Roots: Spreading and woody.
Stems: Tough, woody, thorny.
Leaves: 3 cm long, leathery, dark upper surface, paler underside.
Flowers: Dioecious; male catkins and female flowers without petals.
Seeds: Brown, oblong (6 mm), rich in oils.

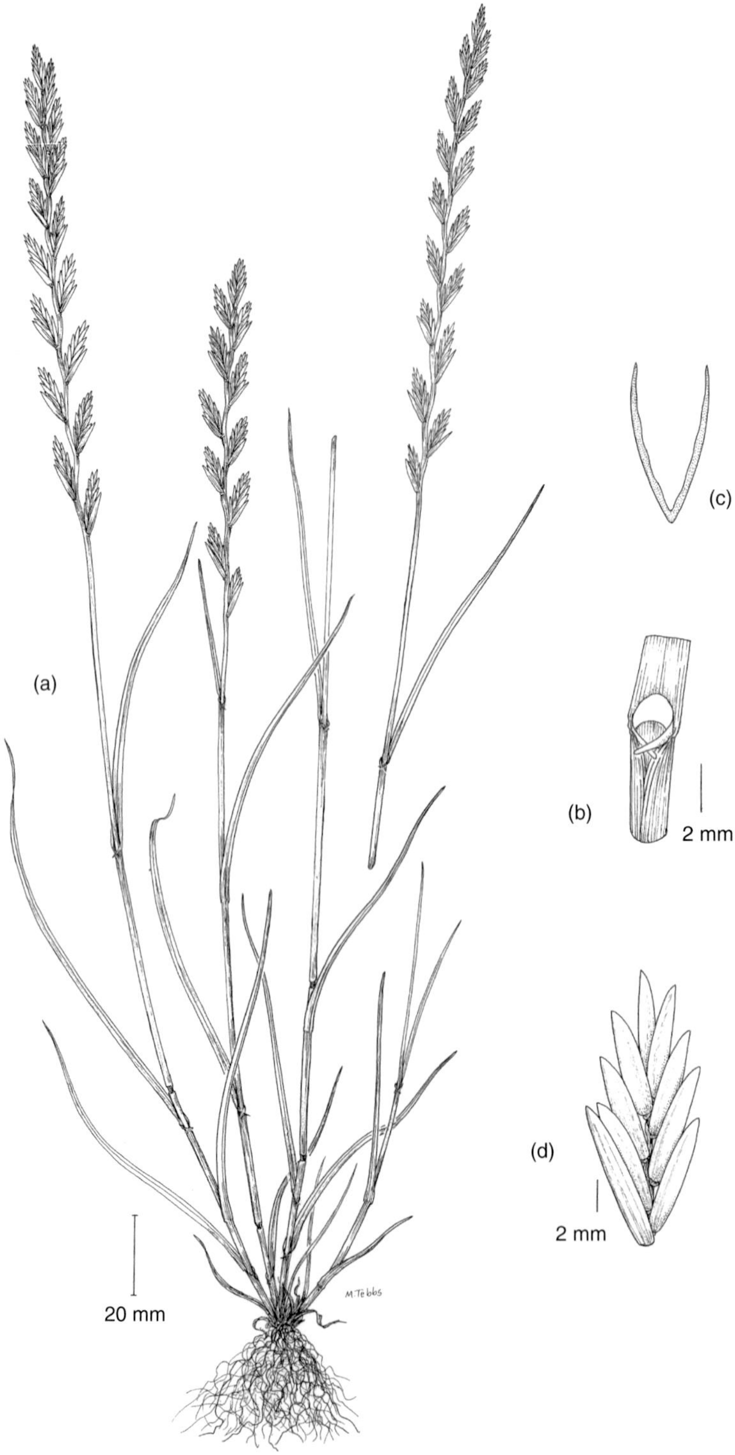

Fig. 7.18. *Lolium perenne* habit (a), ligule (b), spikelet (c) and leaf section (d).

Lolium perenne (Fig. 7.18)

Common name: Perennial ryegrass.
Main uses: Commonly sown as a forage grass; cultivated since the 17th century or earlier.

Environmental benefits

Growing, climate, temperature and rainfall requirements: Vigorous, very hardy, and able to tolerate trampling and mowing. Abundant in old meadows, pastures, parks, road verges and waste ground.
Soil requirements: Adaptable. Will grow in neutral or slightly acidic soils. Frequently sown with clover.
Grazing animals: Modern plant breeders have produced palatable varieties that can withstand heavy grazing.

Structure

Roots: Fibrous.
Stems: Up to 90 cm tall.
Leaves: Youngest blade folded length ways, older blades 2–6 mm wide.
Flowers: Raceme 20–30 cm.
Seeds: Caryopses.

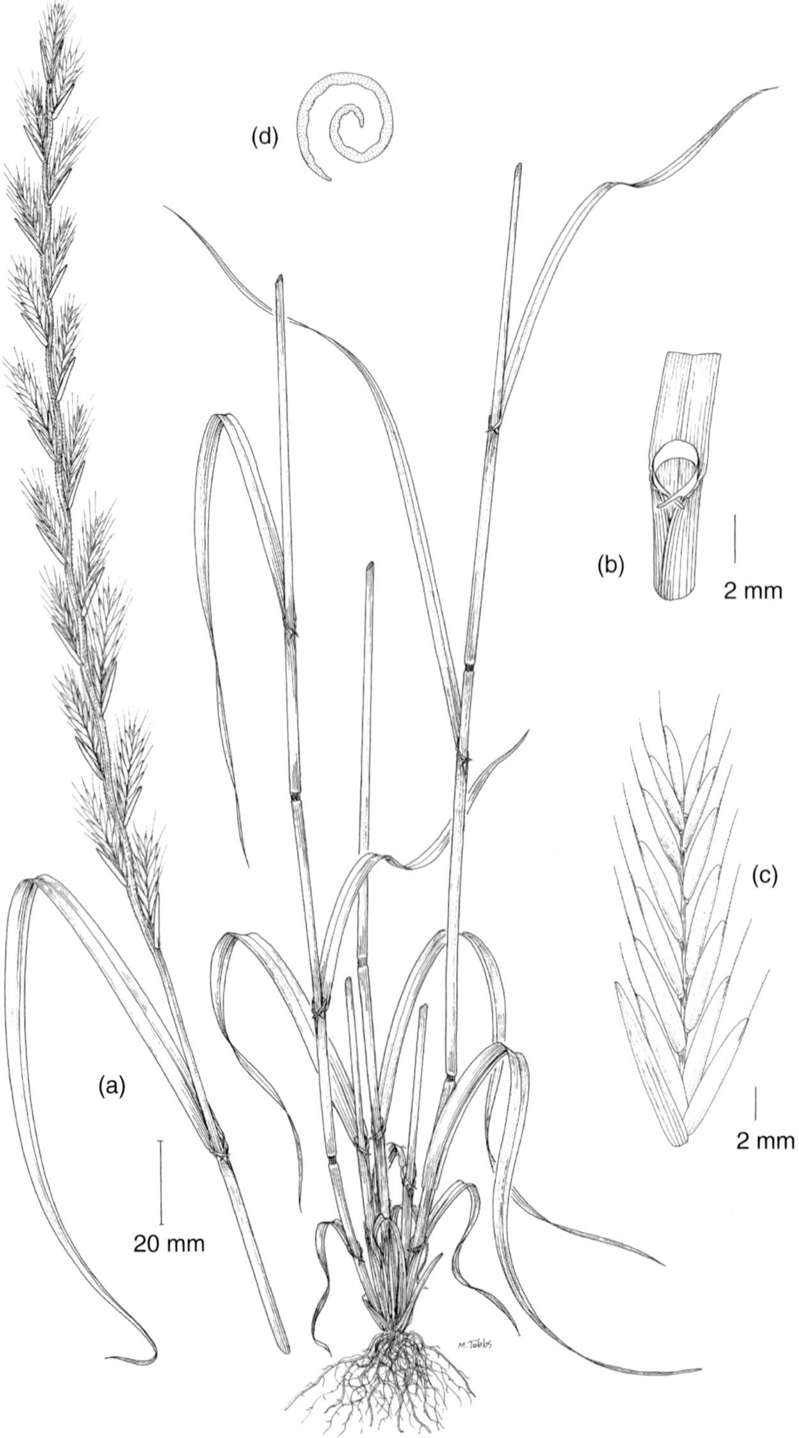

Fig. 7.19. *Lolium multiflorum* habit (a), ligule (b), section of young leaf (c), and spikelet (d).

Lolium multiflorum (Fig. 7.19)

Common name: Italian ryegrass, Westerwold ryegrass.
Main uses: Cover crop; good for building soil structure in orchards, vineyards and crop fields as a mulch. It can reduce soil splash on fruit crops, decrease disease and enhance forage quality.

Environmental benefits

Growing, climate, temperature and rainfall requirements: Annual or occasionally biennial of temperate zones. Outside cultivation, it occurs on disturbed habitats such as roadsides, headlands and waste ground.
Soil requirements: Periodically disturbed ground.
Soil conditioning: Enhances quality of forage and builds up soil structure.
Grazing animals: May be used for fodder.

Structure

Roots: Fibrous.
Stems: Erect.
Leaves: Mature blades flat.
Flowers: Raceme stiff or slightly curved.
Seeds: Caryopses, narrow.

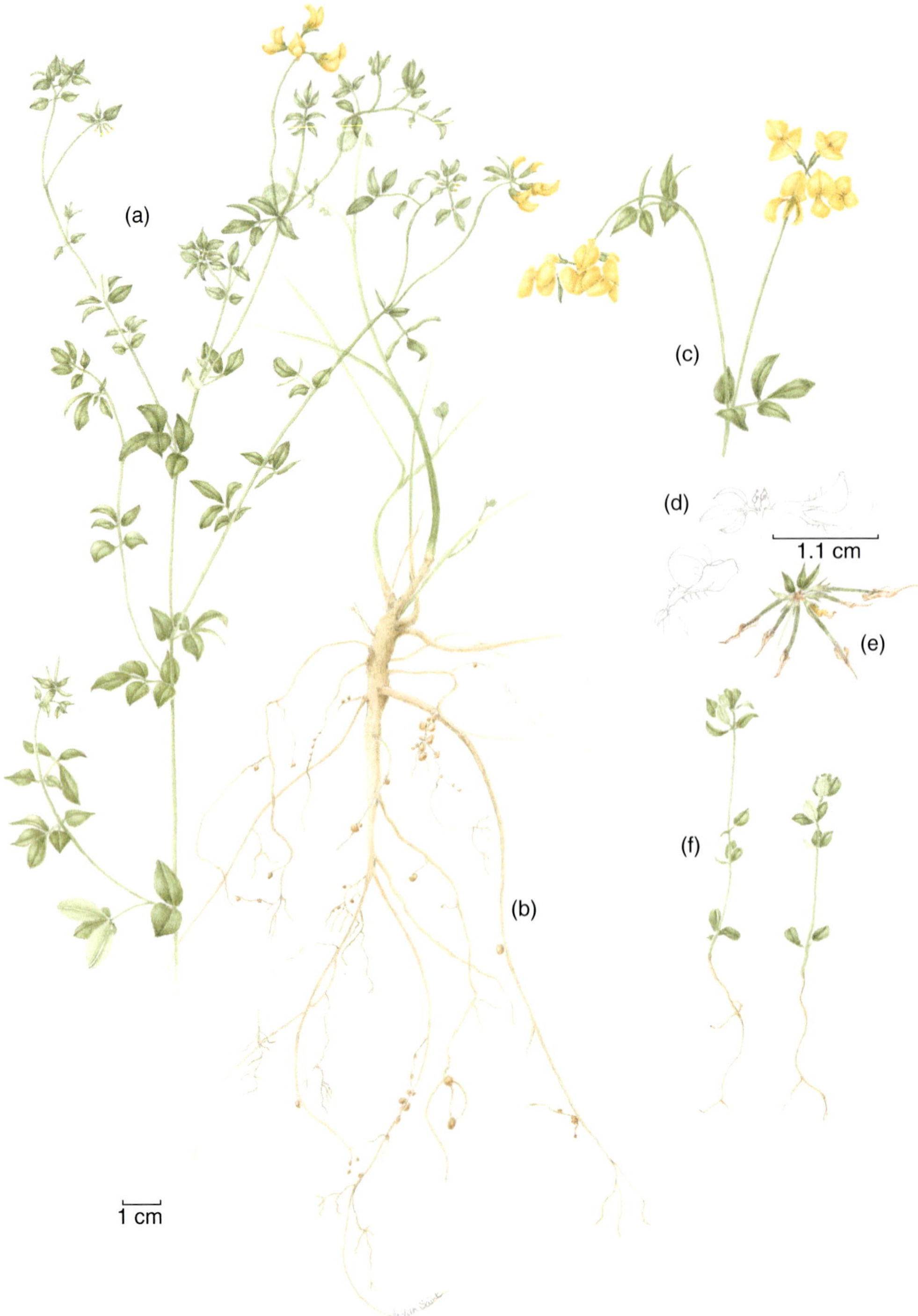

Fig. 7.20. *Lotus corniculatus* habit (a), root (b), flowers (c), flower enlargements (d), seed formation (e) and seedling (f).

Lotus corniculatus (Fig. 7.20)

Common name: Bird's-foot trefoil.
Main uses: Diverse grazing mixtures and environmental schemes.
Unsuitability: Overgrazing will reduce the persistence of bird's-foot trefoil. Leys must have a rest period for recovery.

Environmental benefits

Growing, climate, temperature and rainfall requirements: Winter and drought hardy once established. Best sown in spring. Medium term plants, lasting 2–4 years.
Soil requirements: A wide range of soil types, tolerating a more acidic soil than most legumes. Drought hardy.
Soil conditioning: Fixes nitrogen, improves soil pH and suppresses weeds.
Insects: Many different pollinators visit the flower heads.
Grazing animals: The plants contain condensed tannins, so have anthelmintic and bioactive properties, which improve livestock health, reduce bloat and worms, and allow more efficient processing of protein.

Structure

Roots: It has a deep tap root and also spreads laterally.
Stems: Vertical, spreading, wiry.
Leaves: Small, forming bright green mats.
Flowers: Up to five yellow flowers, sometimes tinged with red, on long stalks.
Seeds: One of the smallest legume seeds. Grow in dark brown, shiny pods in a claw formation.

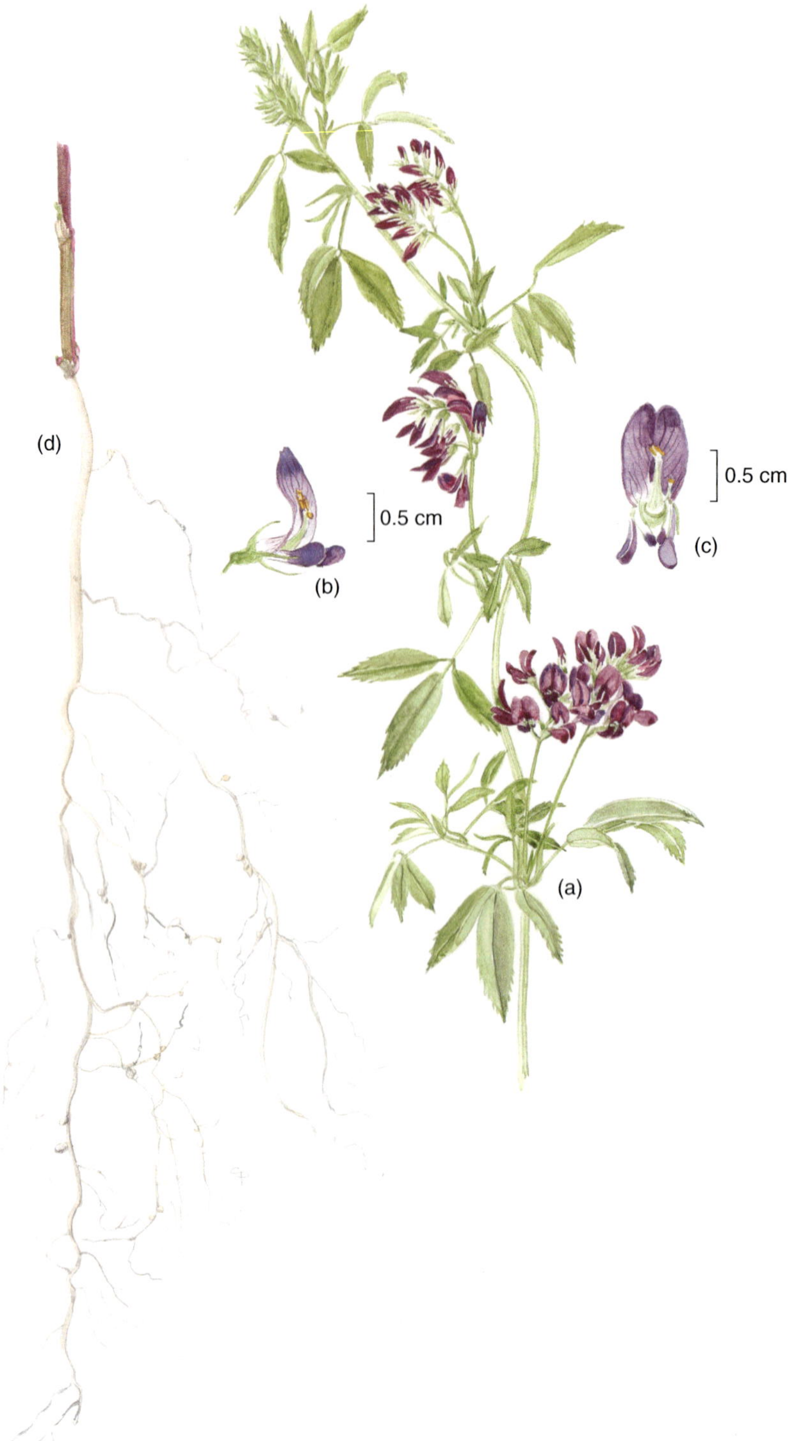

Fig. 7.21. *Medicago sativa* habit (a), side view of flower enlarged (b), front view of flower enlarged (c) and root (d).

Medicago sativa (Fig. 7.21)

Common name: Lucerne or alfalfa.
Main uses: Hay crop, silage, green manure.
Unsuitability: Unsuitable for poorly drained or acidic soils.

Environmental benefits

Growing, climate, temperature and rainfall requirements: Sow in warm spring soil, at 7°C or more. Late summer before the soil cools is best for autumn sowing. Winter hardiness varies with variety. Be sure to get a variety best suited to local winter temperatures. Drought tolerant.
Soil requirements: Needs a naturally free-draining soil with pH 6.2 or above. Cannot tolerate acidic or waterlogged soil.
Soil conditioning: Fixes nitrogen once a compatible *Rhizobium* association is established. Can be inoculated. Its abundant mass makes good green manure.
Insects: Good resource for pollinators.
Grazing animals: Should only be grazed lightly at the end of the first year. Thereafter, back-fenced grazing prevents both overgrazing on the sward and bloat in the animals. Recovery time between grazing should be allowed.

Structure

Roots: Very deep tap root.
Stems: Upright and green.
Leaves: Tapering, hairless, trifoliate.
Flowers: Purple.
Seeds: Smooth, beige to brown, and oblong or kidney shaped.

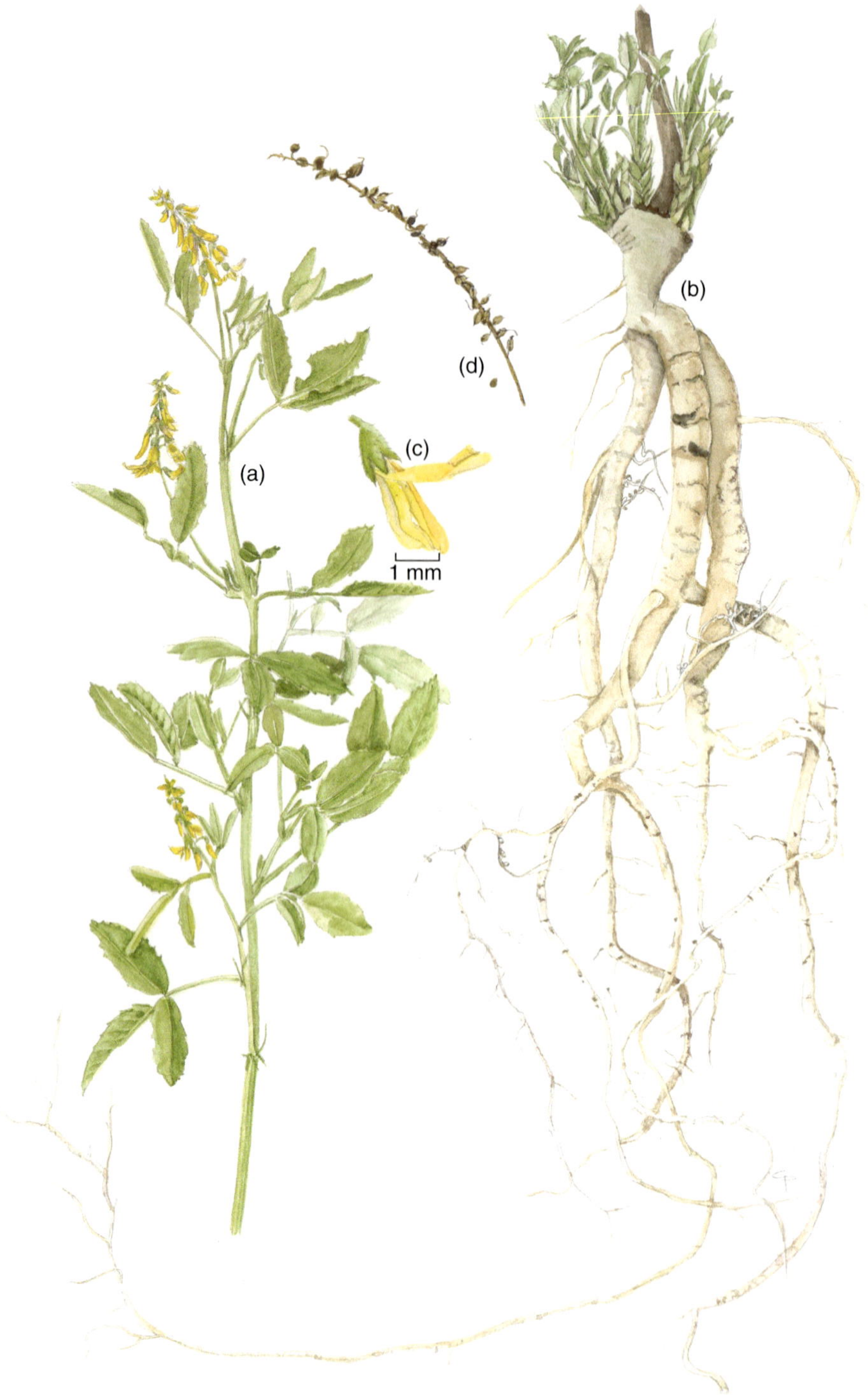

Fig. 7.22. *Melilotus officinalis* habit (a), root (b), flower enlargement (c) and seeds (d).

Melilotus officinalis (Fig. 7.22)

Common name: Sweet clover.
Main uses: Diverse mixtures, cover crops, game cover.
Unsuitability: Deep roots can make it difficult to terminate volunteers in immediate subsequent crops.

Environmental benefits

Growing, climate, temperature and rainfall requirements: Sow in warm spring soil but not in the autumn. Winter hardy as a crown and root, sprouting in the spring. Drought tolerant.
Soil requirements: Wide range of soil types. May benefit from *Rhizobium* inoculation in less fertile soils.
Soil conditioning: Fixes atmospheric nitrogen and develops strong deep roots, which improve drainage and break up soil compactions.
Insects: Good resource for pollinators and other beneficial insects.
Grazing animals: Should be in a mixed sward to prevent bloat.

Structure

Roots: Deep, branched tap roots and smaller laterals.
Stems: Become dull in colour and woody in the second year. Up to 1.5 m tall.
Leaves: Trifoliate, oblong.
Flowers: Racemes may be yellow or white.
Seeds: Rounded, oblong, in pods.

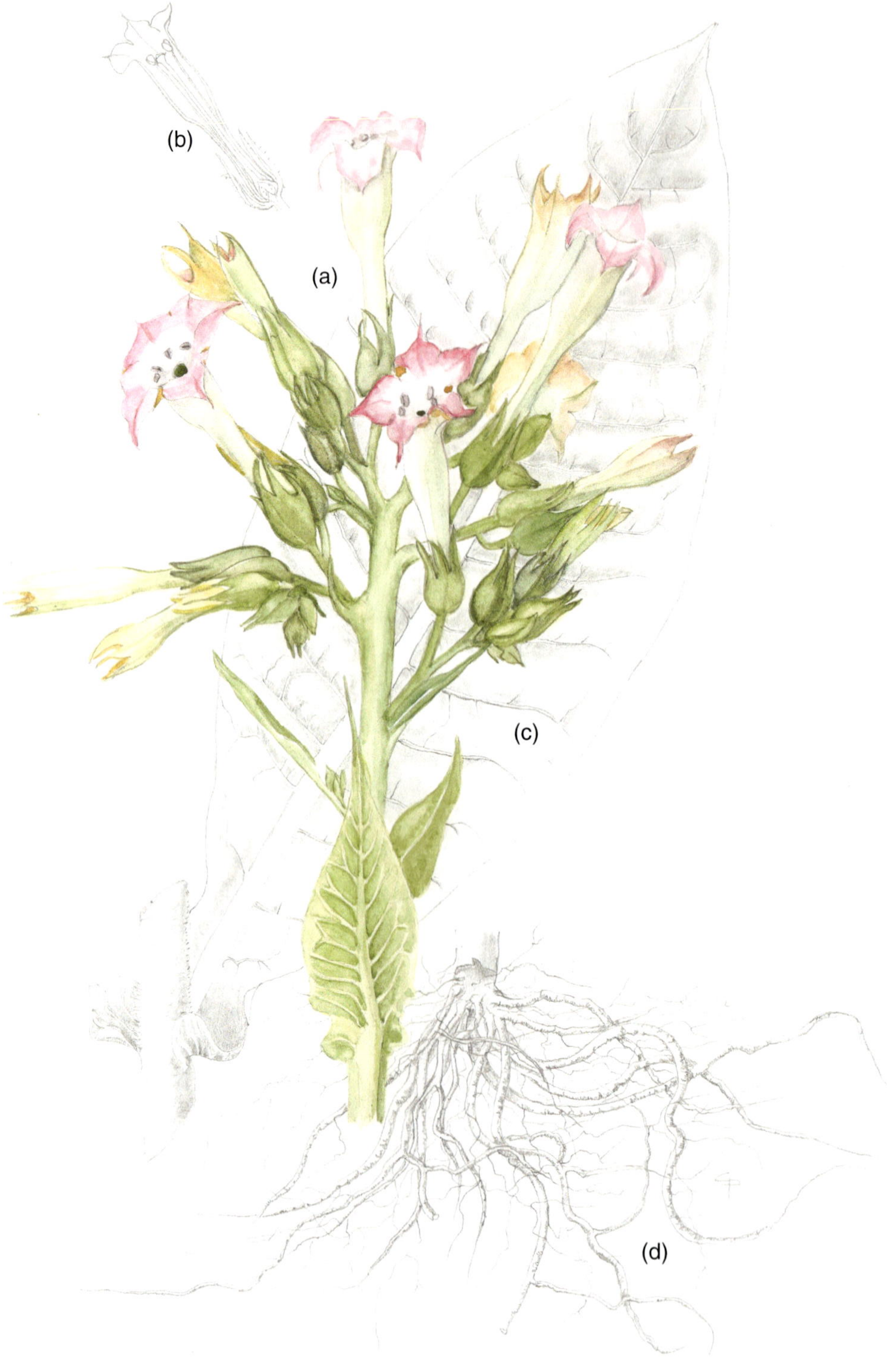

Fig. 7.23. *Nicotiana tabacum* flower (a), flower vertical section (b), leaf (c) and root (d).

Nicotiana tabacum (Fig. 7.23)

Common name: Tobacco.
Main uses: Commercial tobacco for smoking, phytoremediator.
Unsuitability: It has no food use and does not thrive in cooler climates. It is associated with high use of agrichemicals. Its relatively shallow rooting makes it suitable for cleaning soils only where the contaminants are near the surface.

Environmental benefits

Growing, climate, temperature and rainfall requirements: Seed is scattered on the surface and then protected until the seedlings mature. Moisture and light are needed for germination, but the mature plant can tolerate aridity. Not frost hardy.
Soil requirements: Moist well-drained soils are ideal.
Soil conditioning: Tobacco absorbs both organic and inorganic pollutants, cleaning soils of cadmium, lead, mercury, arsenic and uranium. Organic pollutants remediated include different hydrocarbons, chlorinated solvents and trinitrotoluene (TNT).
Insects: Insect pollinated. Extensive monoculture acreages attract many insect predators.
Grazing animals: Not grazed.

Structure

Roots: Fibrous, shallow, branched.
Stems: Erect, becoming woody as the seeds mature.
Leaves: Large, oblanceolate.
Flowers: Pink, white or creamy trumpet-shaped flowers.
Seeds: Small (1 mm), brown, broadly ovate.

Fig. 7.24. *Onobrychis viciifolia* habit (a), vertical section of flower enlarged (b), flower enlarged (c), root (d), root enlargement (e), seed (f) and seed enlarged (g).

Onobrychis viciifolia (Fig. 7.24)

Common name: Sainfoin.
Main uses: Low-input forage crop. Nitrogen fixer. Can be grown as a monocrop for hay, or sown or overseeded into a paddock for a mixed sward.
Unsuitability: Should not be grazed too heavily. Slower to establish than some clovers. Does not do well where chemical nitrogen is applied.

Environmental benefits

Growing, climate, temperature and rainfall requirements: Ideally deep sown in late spring in damp soil and rolled after sowing to increase seed contact with moisture. Frost hardy and drought tolerant. Can be overseeded into a paddock for mixed grazing.
Soil requirements: A wide range of soil types, flourishing on poor soils. Particularly likes calciferous soils.
Soil conditioning: If left as a ley for 3–4 years, its deep roots reduce soil compaction, and the optimum carbon:nitrogen ratio is perfect for soil organisms. Fixes slightly less nitrogen than red clover.
Insects: Loved by pollinators and improves wildlife diversity.
Grazing animals: The high tannin and polyphenol content makes it a potential alternative to synthetic anti-parisitic drugs. The tannins also reduce the incidence of greenhouse gas emissions from ruminants.

Structure

Roots: Deep tap root.
Stems: Strong and erect, growing to 75 cm.
Leaves: Pinnate compound leaf.
Flowers: Bright pink, cone-shaped, erect panicles.
Seeds: Large (1 cm), flat and semi-circular with horned ridges.

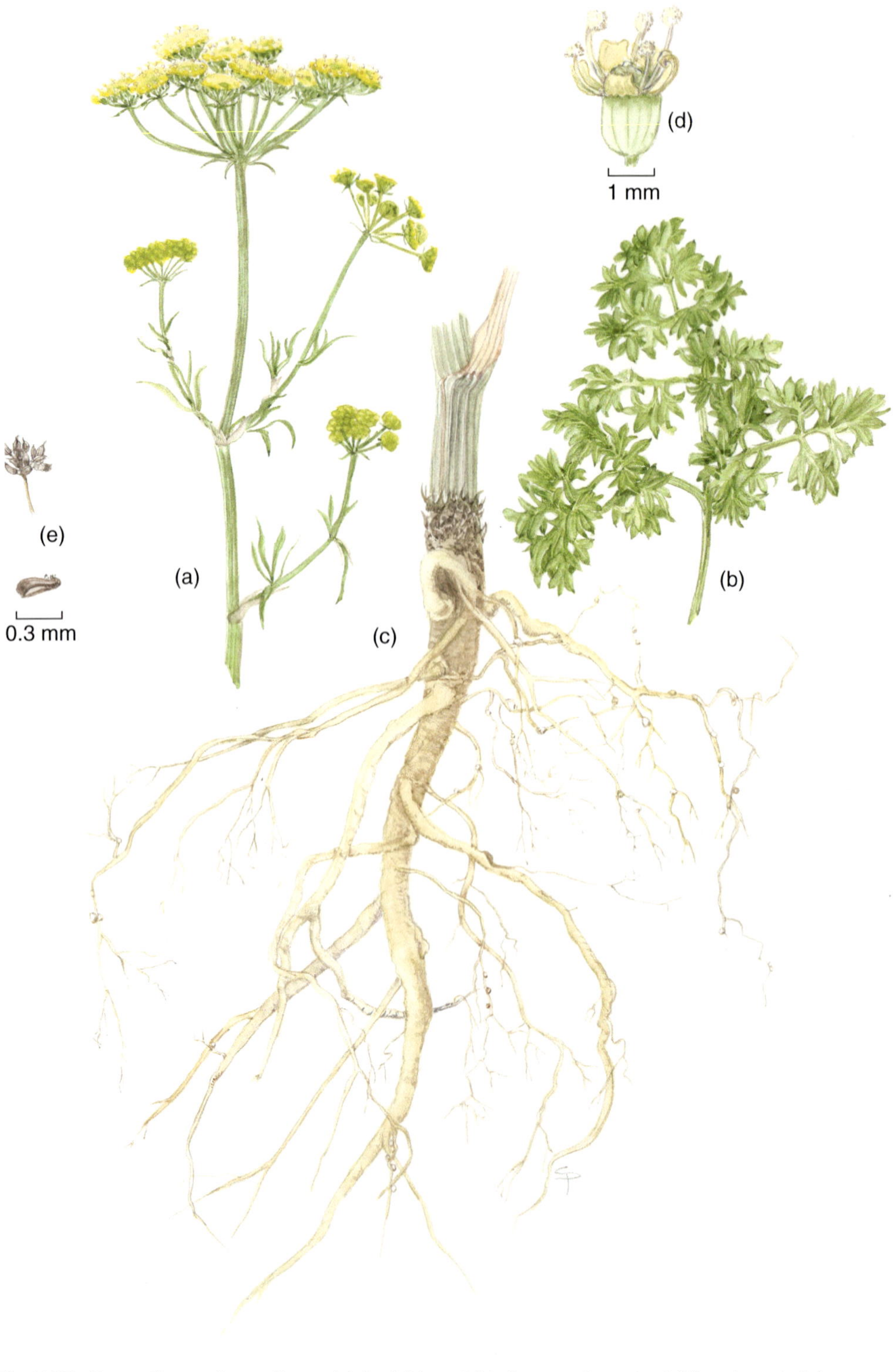

Fig. 7.25. *Petroselinum crispum* flower (a), leaf (b), root (c), flower enlargement (d) and seeds (e).

Petroselinum crispum (Fig. 7.25)

Common name: Sheep's parsley.
Main uses: Forage herb mixed with ribgrass, yarrow and burnet.
Unsuitability: Not suitable for heavy grazing.

Environmental benefits

Growing, climate, temperature and rainfall requirements: The small seeds must be sown at a shallow depth in spring and autumn. Frost tolerant.
Soil requirements: A wide range of soil types.
Soil conditioning: The deep roots are good for drought tolerance and soil conditioning.
Insects: Good resource for pollinators and other beneficials.
Grazing animals: Not tolerant to tramping or heavy grazing, but has medicinal qualities for animals, being high in iron and vitamins A and C, enhancing fertility in breeding stock.

Structure

Roots: Deep, branched tap roots and smaller laterals.
Stems: Become dull in colour and woody in the second year. Up to 1 m tall.
Leaves: Similar to parsley, deeply incised.
Flowers: Yellow umbels.
Seeds: Oblong and ridged.

Fig. 7.26. *Phacelia tanacetifolia* habit (a), root (b), flower enlargement (c) and seedling (d).

Phacelia tanacetifolia (Fig. 7.26)

Common name: Phacelia.
Main uses: Quick-growing, weed-suppressing foliage. Good for pollinators.

Environmental benefits

Growing, climate, temperature and rainfall requirements: Shallow sowing at any time from early spring. Will overwinter in a mild climate.
Soil requirements: Will grow in most soils, but particularly in dry ones.
Soil conditioning: Produces good biomass quickly and will break down rapidly after mulching.
Insects: Good resource for pollinators and other beneficials.
Grazing animals: Not normally used in pasture.

Structure

Roots: Shallow and branched.
Stems: Up to 1.5 m tall.
Leaves: Fine and feathery.
Flowers: Prolific and blue.
Seeds: Oblong, black.

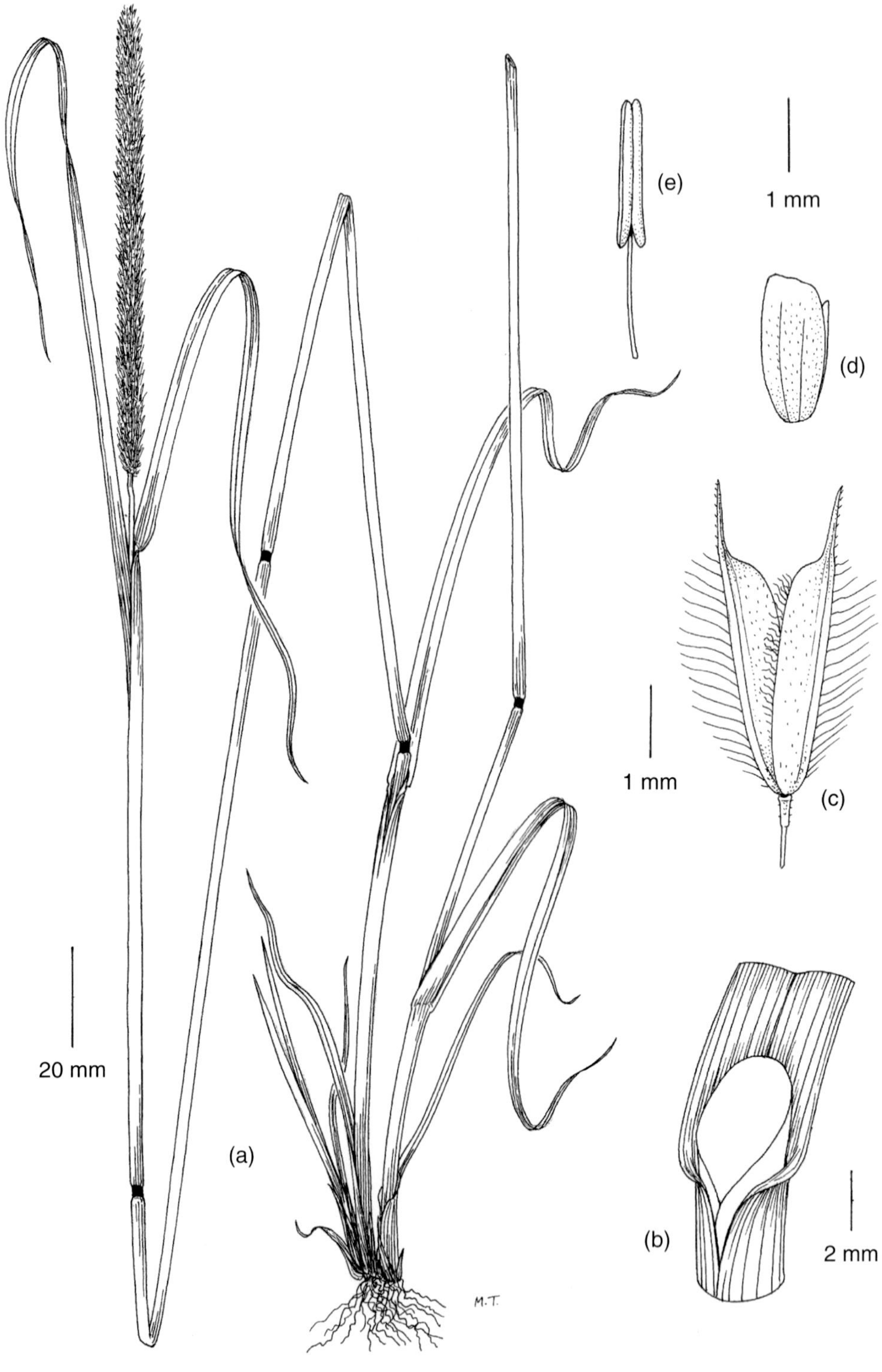

Fig. 7.27. *Phleum pratense* habit (a), ligule (b), glumes (c), floret (d) and anther (e).

Phleum pratense (Fig. 7.27)

Common name: Timothy.
Main uses: Hay or grazing.

Environmental benefits

Growing, climate, temperature and rainfall requirements: Vigorous, very hardy. Able to tolerate cold. Stays green in winter.
Soil requirements: Moist, heavy or peaty soils.
Soil conditioning: Widely grown for hay, silage and pasture. A variable species with many cultivars.
Grazing animals: Vulnerable to heavy grazing.

Structure

Roots: Fibrous.
Stems: Up to 150 cm tall.
Leaves: Narrow, mostly glabrous, greyish-green.
Flowers: Cylindrical panicle up to raceme, up to 14 cm.
Seeds: Caryopses.

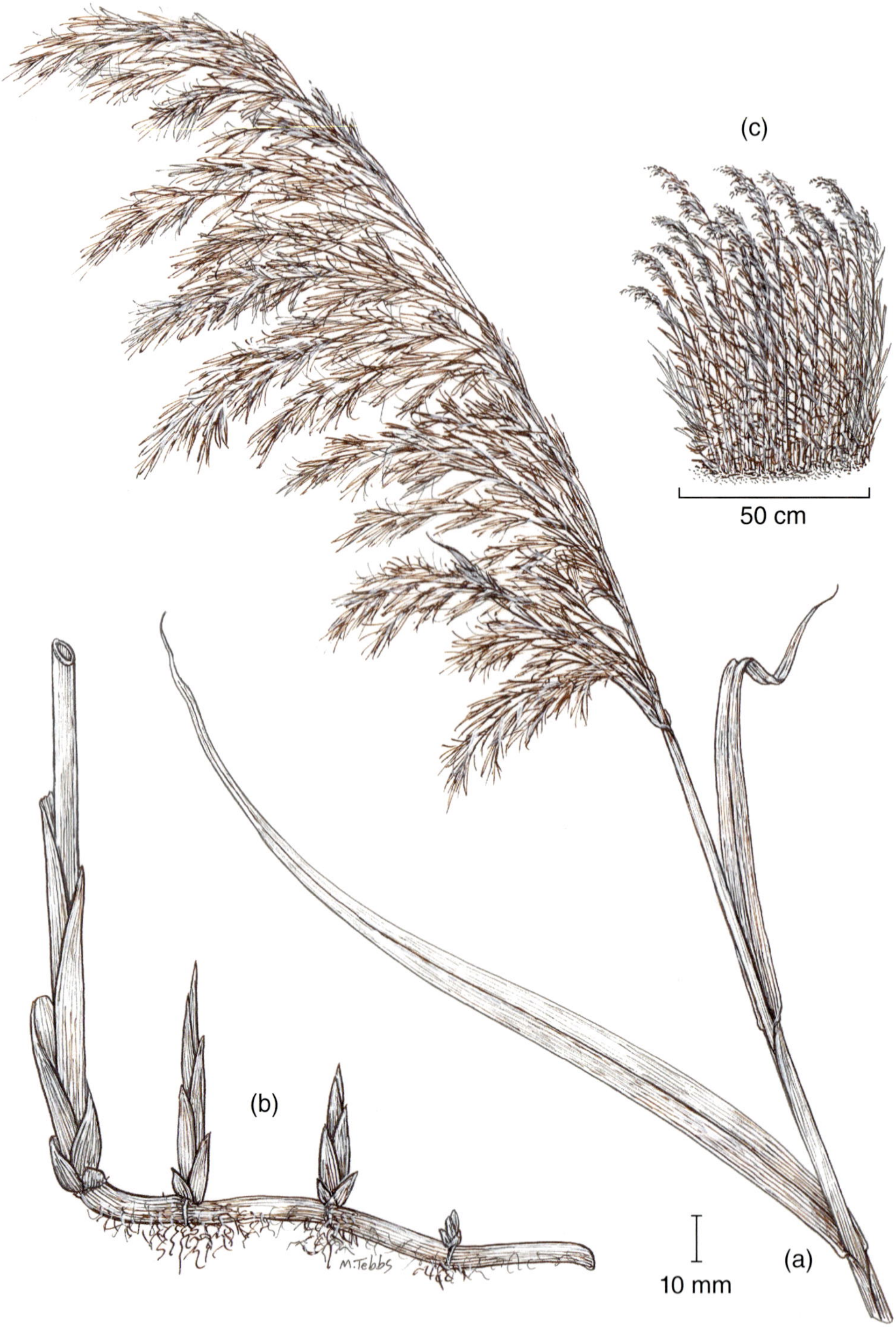

Fig. 7.28. *Phragmites australis* habit (a), root (b) and stand of reeds growing in mud (c).

Phragmites australis (Fig. 7.28)

Common name: Common reed.
Main uses: Permaculture clean-up system for water treatment and sewage effluent, domestic thatching, stabilizing land and preventing erosion.
Unsuitability: Not suitable for cultivation in gardens.

Environmental benefits

Growing, climate, temperature and rainfall requirements: Perennial of wetland habitats, ditches, marshes, swamps and fens. Temperate, circumpolar climates. It can grow rapidly, forming large invasive stands.
Soil requirements: Moist or waterlogged. Grows by lakes and brackish water.
Soil conditioning: Can transfer oxygen through stems into areas around the root system, creating high populations of microorganisms.
Grazing animals: It is an invasive plant, and can be kept down by grazing of cattle and goats. It also provides excellent habitat for marsh birds.

Structure

Roots: Extensive system of rhizomes, which may penetrate over 1 m deep.
Stems: Up to 3.5 m tall.
Leaves: 20–50 cm long, curved and fragile.
Flowers: Inflorescence is a loose, soft, dense panicle.
Seeds: Mostly vegetative; seed set is erratic.

Fig. 7.29. *Plantago lanceolata*.

Plantago lanceolata (Fig. 7.29)

Common name: Ribgrass.
Main uses: Pasture mixtures. Useful ground cover in vineyards.
Unsuitability: The pollen can cause allergies and respiratory problems.

Environmental benefits

Growing, climate, temperature and rainfall requirements: Flood, drought and frost resistant but not invasive. Grows everywhere except arctic and lowland tropical areas.
Soil requirements: Wide range of soil types. Found in all but the most acidic grassland on all continents.
Soil conditioning: Mineral rich – a well-developed tap root brings trace elements to the surface.
Insects: Although largely wind pollinated, it also hosts small bees, flies and insects that feed on the mildew to which it is subject. The moths hosted are barred chestnut, Hebrew character, golden dart and striped rustic.
Grazing animals: Its high protein content (~20%) helps the production of milk and meat products. Contains tannins and so is medicinal to animals, having anthelmintic, anti-inflammatory, antihistamine, antifungal, antioxidant, analgesic and even mild antibiotic properties. It is especially palatable to sheep.

Structure

Roots: Well-developed tap roots.
Stems: Erect, hairy, square in profile and deeply furrowed; bend over as seeds develop.
Leaves: Basal rosette leaves with wide variation of width depending on location, but always with parallel veins, smooth and broadly lanceolate in shape.
Flowers: Creamy white racemes.
Seeds: Capsules 3–4 mm long containing one to two mucilaginous seeds.

Fig. 7.30. *Populus tremula* tree (a), bark (b), leaf (c) and petiole (d).

Populus tremula (Fig. 7.30)

Common name: European aspen.
Main uses: Phytoremediator and phytostabilizer, reducing pollutants over time. Can grow in many places contaminated by slag heaps, heavy metals and smelter pollution. Aspen has colonized large areas of polluted industrial sites, containing poisonous residues in its roots.
Unsuitability: Not recommended for suburban gardens, as it is prone to spreading by suckering.

Environmental benefits

Growing, climate, temperature and rainfall requirements: Fast-growing deciduous plant. Grows in cool conditions.
Soil requirements: Moist, often acid soil.
Grazing animals: Wild animals such as deer and moose will eat the bark in cold seasons.

Structure

Roots: Invasive, spreading rapidly across a wide area and forming suckers. Can cause damage in gardens to concrete and brickwork.
Stems: Dark brown, slender, shiny.
Leaves: Ovate to rounded with large uneven coarse teeth and long flattened petioles.
Flowers: The trees are dioecious. Male catkins with red anthers, female catkins with many ovaries.
Seeds: Small and fluffy. However, mostly growth is vegetatively from root suckers.
Similar species: *Populus tremuloides* (North American aspen), which differs by having finely toothed leaves. Sometimes hybridizes with *P. tremula*.

Fig. 7.31. *Prunus domestica* subsp. *insititia* (top) and *Prunus cerasifera* (bottom).

Prunus domestica subsp. *insititia* and *Prunus cerasifera* (Fig. 7.31)

Common names: Damson and myrobalan red plum.
Main uses: Hedging windbreaks and culinary fruit crop. Used as root stock for grafting cultivars.
Unsuitability: Myrobalan red plum is an erratic fruiter in years of heavy frost. Young trees take a long time to start producing fruit.

Environmental benefits

Growing, climate, temperature and rainfall requirements: Flower in April in the northern hemisphere, and fruit from August to September. Robust and hardy. For reliable cropping of myrobalan red plum, areas prone to late frosts should be avoided. Damsons are winter hardy and can be grown as far north as Norway. They need moisture at depth throughout the summer. Can be propagated from suckers and seed, although the seedlings may not come true.
Soil requirements: Wide range of soil types but need a good depth of well-drained soil. A thin covering of soil over chalk is not suitable.
Soil conditioning: Deep roots help drainage and soil compaction.
Insects: Insect pollinated.
Grazing animals: Not normally.

Structure

Roots: Deep and strong.
Stems: Erect, becoming woody as the seeds mature.
Leaves: Entire, toothed.
Flowers: Small white clusters.
Seeds: Deeply furrowed and clinging to the fruit flesh.

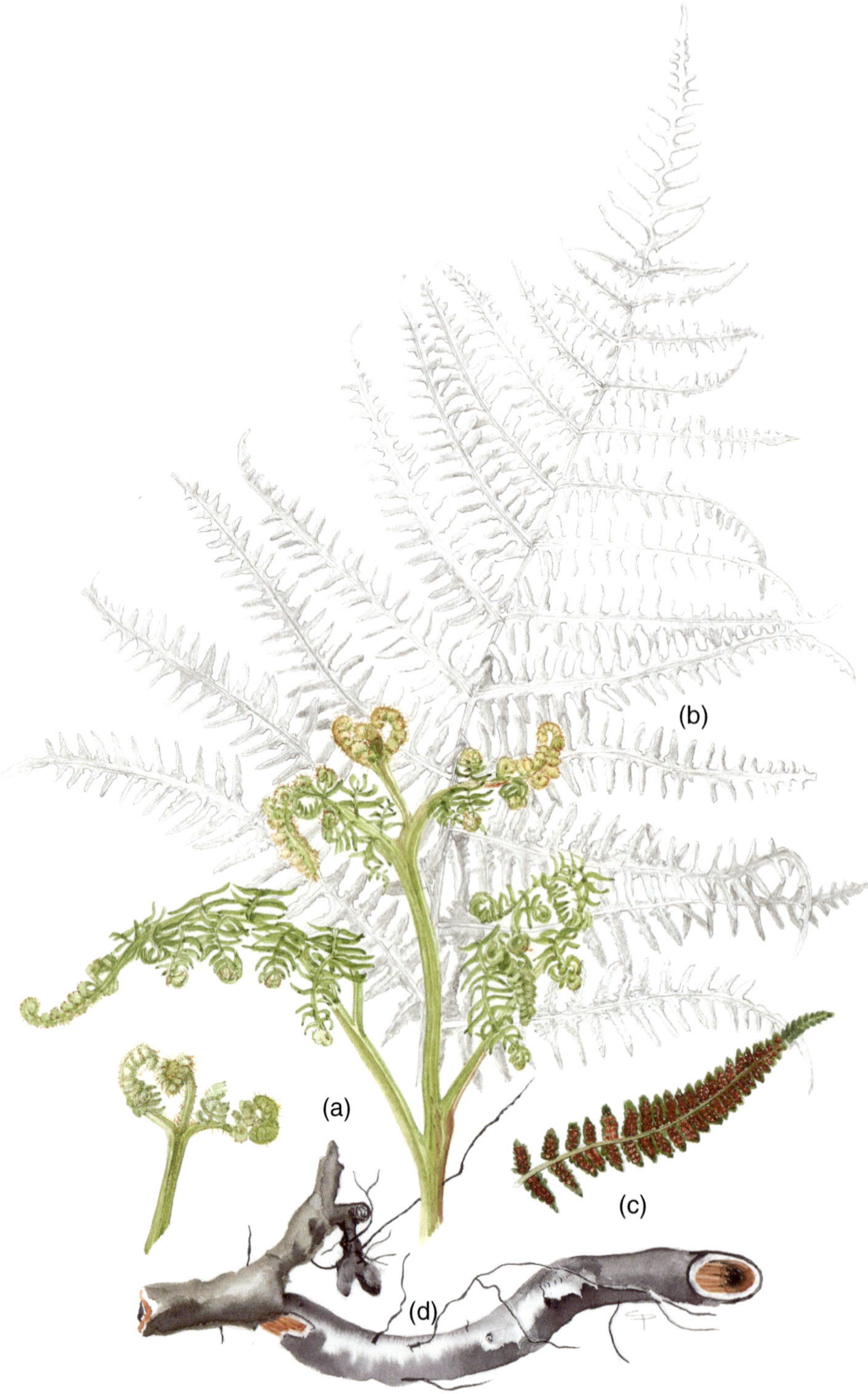

Fig. 7.32. *Pteridium aquilinum* young fronds (a), mature fronds (b), spores (c) and root (d).

Pteridium aquilinum (Fig. 7.32)

Common name: Bracken.
Main use: Game cover, potential biofuel or mulch.
Unsuitability: An invasive plant dominating wide areas of lightly managed land.

Environmental effects

Growing, climate, temperature and rainfall requirements: Tender to a late spring frost, salt spray and wind. Drought tolerant.
Soil requirements: Prefers well-drained acid soils that are not compacted. It will not spread into waterlogged areas.
Insects: Some moth larvae feed on bracken. It is the main food plant for bracken neb (*Monochroa cytisella*), brown silver-line (*Petrophora chlorosata*) and *Callopistria juventina* moth larvae.
Grazing animals: Toxic to all animals, especially the young shoots. Animals avoid eating it if there is anything else. Sheep grazing in areas of encroachment encourage spread by reducing competition.

Structure

Roots: Strong rhizomes.
Stems: Tough, crescent-shaped in cross-section.
Leaves: Pinnate, in a fractal pattern.
Flowers: None.
Seeds: Spores form on the undersides of some mature fronds.

Fig. 7.33. *Raphanus sativus* and *R. raphanistrum* habit (a), root (b), flower enlargement (c) and mature leaf (d).

Raphanus sativus and *R. raphanistrum* (Fig. 7.33)

Common name: Tillage, fodder, forage or oil radish (*R. sativus*) and wild radish (*R. raphanistrum*).
Main uses: Winter ground cover, mob grazing, fodder crop.
Unsuitability: Not suitable for an overwinter crop in colder climates where night temperatures frequently drop below freezing.

Environmental benefits

Growing, climate, temperature and rainfall requirements: Sown in early autumn for an overwintering crop, but can be sown from late spring onwards for quicker growth.
Soil requirements: All types of soil.
Soil conditioning: The long tap roots penetrate deep into the soil, draw up nutrients and improve soil structure, moisture-holding capacity and fertility.
Grazing animals: Ideal for mob grazing sheep at the end of winter. Will regrow for a second crop, followed by turning in the roots.

Structure

Roots: Fleshy tap roots.
Stems: Become dull in colour and woody in the second year. Up to 1.5 m tall.
Leaves: Lobed.
Flowers: White or yellow.
Seeds: Oblong, 3 mm long, in pods.

Fig. 7.34. *Reynoutria japonica* flowering stem (a) and single flower enlargement (b).

Reynoutria japonica (Fig. 7.34)

Common name: Japanese knotweed.
Main uses: Contains the antibiotic resveratrol. Provides a nectar source in late summer.
Unsuitability: Invasive and tenacious.

Environmental benefits

Growing, climate, temperature and rainfall requirements: Common in temperate climates, robust.
Soil requirements: Tolerant of a wide range of soil types but likes moist soils.
Insects: Good resource for pollinators and other beneficials.
Grazing animals: Grazed by goats and pigs. Heavy grazing helps to control growth.

Structure

Roots: Deep rhizomes up to 3 m in depth.
Stems: Hollow and jointed.
Leaves: Broad and heart-shaped.
Flowers: Sprays of racemes of small white flowers.
Seeds: Winged nut 5 mm long. Eaten by birds.

Fig. 7.35. *Rumex obtusifolius* flower (a), leaf (b) and root (c).

Rumex obtusifolius (Fig. 7.35)

Common name: Broad-leaved dock.
Main uses: Component of grazing pastures, reducing bloat in leguminous forage.
Unsuitability: Deep roots and robust survival strategies make it difficult to eradicate.

Environmental benefits

Growing, climate, temperature and rainfall requirements: Tolerant of wide ranges in temperature. Drought tolerant, flood tolerant and frost hardy.
Soil requirements: Wide range of soil types, preferring clay, chalk and gravel to light sandy soils.
Soil conditioning: The strong, deep roots improve drainage and break up soil compactions.
Insects: Host to several species of insects.
Grazing animals: Sheep, goats, cows and horses graze successfully, with dock adding valuable trace elements to the forage diet.

Structure

Roots: Deep, branched tap roots and smaller laterals.
Stems: Become tough with maturity.
Leaves: Elliptic.
Flowers: Spikes of green and white, turning crimson.
Seeds: 5–6 mm, winged.

Fig. 7.36. *Salix viminalis* leaf (a), new growth (b) and catkin enlargements (c): female (1) and male (2).

Salix viminalis (Fig. 7.36)

Common name: Osier willow.
Main uses: Phytoremediation, soil stabilization and erosion control, woody biomass fuel, game cover, basket making.
Unsuitability: Considered an invasive species in temperate parts of Australia.

Environmental benefits

Growing, climate, temperature and rainfall requirements: Cold temperate regions. Fast growing and easily established from unrooted cuttings, planted in weed-free soil during the winter. Osier willow can be coppiced in alternate years.
Soil requirements: Marginal and degraded soil, landfill sites or soil subject to flooding. Sandy and peaty soils are less successful.
Soil conditioning: Low maintenance and cost effective, osier willow has efficient nutrient and nitrogen uptake. It effectively filters wastewater, biosolids and soil contaminants. It replaces contaminants with leaf litter over time.
Insects: Early catkins are a good nectar source for bees. Butterflies and moths lay their eggs on osier willow, which supports a large number of insects without suffering material damage.
Grazing animals: Goats and horses browse on osier willow, and it is a useful winter cut fodder crop for cattle and sheep.

Structure

Roots: Fine white roots develop into a shallow network.
Stems: Woody.
Leaves: Slender and pointed, up to 7 cm long.
Flowers: The trees are dioecious, with male and female catkins on separate plants.
Seeds: Fluffy, wind dispersed.

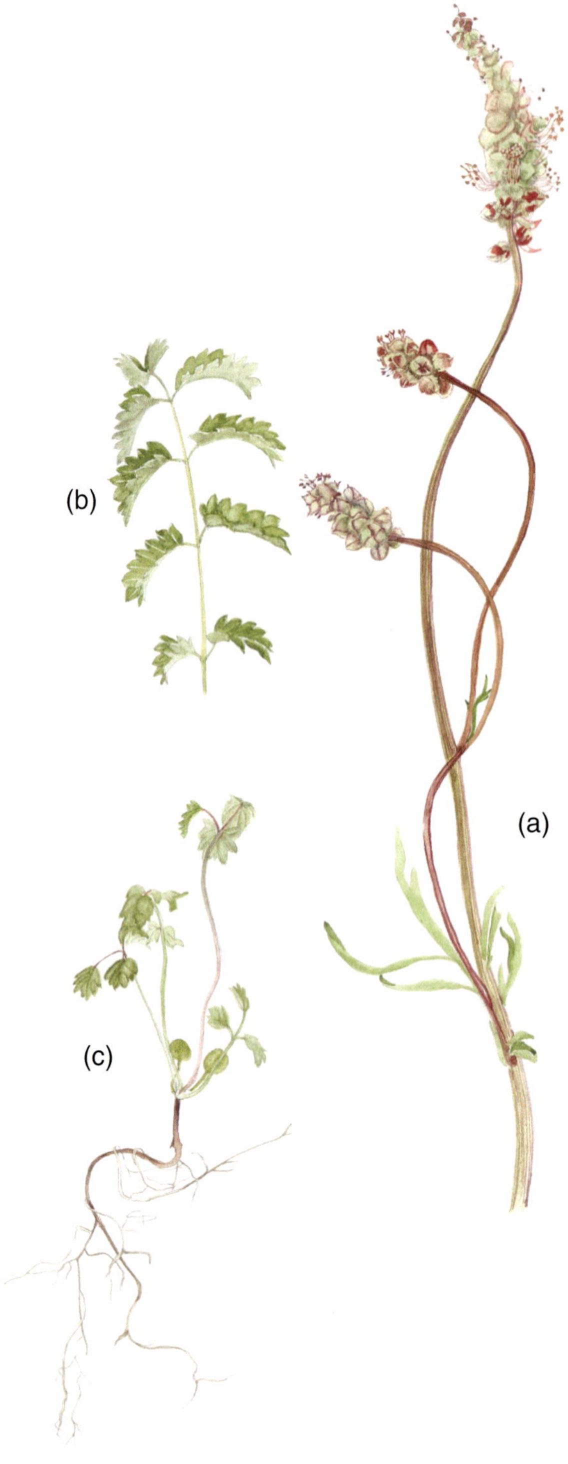

Fig. 7.37. *Sanguisorbe minor* flower (a), leaf (b) and seedling (c).

Sanguisorbe minor (Fig. 7.37)

Common name: Salad burnet.
Main uses: Forage, ground cover, erosion control.
Unsuitability: Not suitable as a monocrop or for overgrazing when young.

Environmental benefits

Growing, climate, temperature and rainfall requirements: Perennial. Grows in early spring for a late spring forage. Good frost and drought tolerance.
Soil requirements: Prefers damp but drained soils with a pH of 6.8. Tolerant of thin soils.
Soil conditioning: The long roots lift trace elements from deep below the surface.
Insects: Food plant for the larvae of the grizzled skipper and mouse moth.
Grazing animals: Valuable component of mixed forage for persistence and minerals.

Structure

Roots: Long.
Stems: Become woody with age.
Leaves: Alternate, serrated, pinnate.
Flowers: Small, tight clusters of red to white four-petalled flowers from June to September.
Seeds: Oval in shape and angular.
Similar species: 30 species; the genus hybridizes readily, producing numerous cultivars.

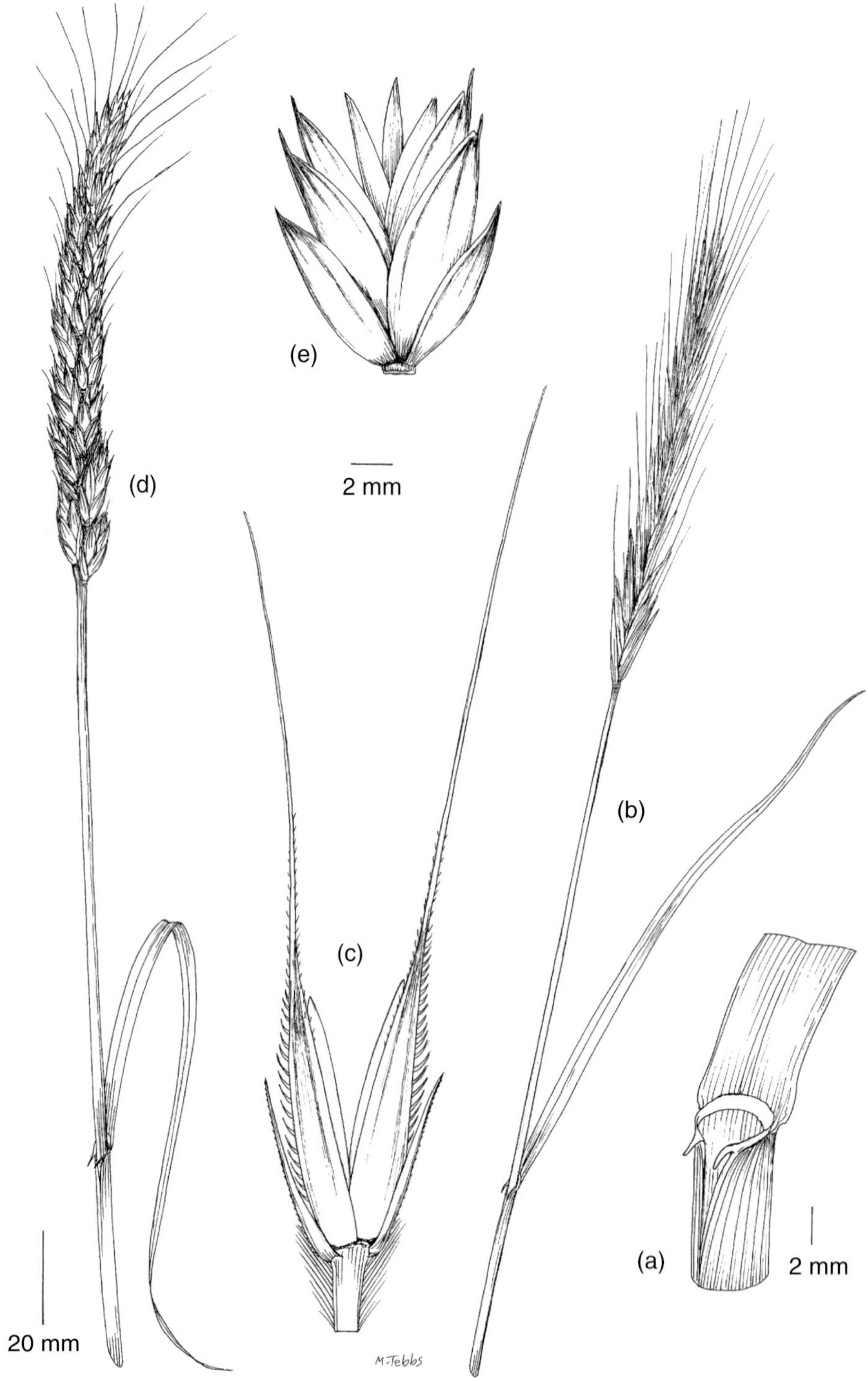

Fig. 7.38. *Secale cereal* ligule (a), inflorescence (b), spikelet (c) and ×*Triticosecale* inflorescence (d) and spikelet (e).

Secale cereal (Fig. 7.38)

Common name: Rye.
Main uses: Cover crop, green manure and early pasture. It is a grain crop in some areas, used for black bread or malt, and may be cut for domestic thatching.

Environmental benefits

Growing, climate, temperature and rainfall requirements: Vigorous, very hardy and able to tolerate cold and drought.
Soil requirements: Sandy infertile soils.
Soil conditioning: Decomposing residue of winter rye has been shown to chemically suppress germination of weed seeds.
Grazing animals: Useful grass for early pasture.

Structure

Roots: Fibrous.
Stems: Up to 150 cm tall.
Leaves: Narrow, mostly glabrous blades.
Flowers: Racemes up to 20 cm long.
Seeds: Caryopses.

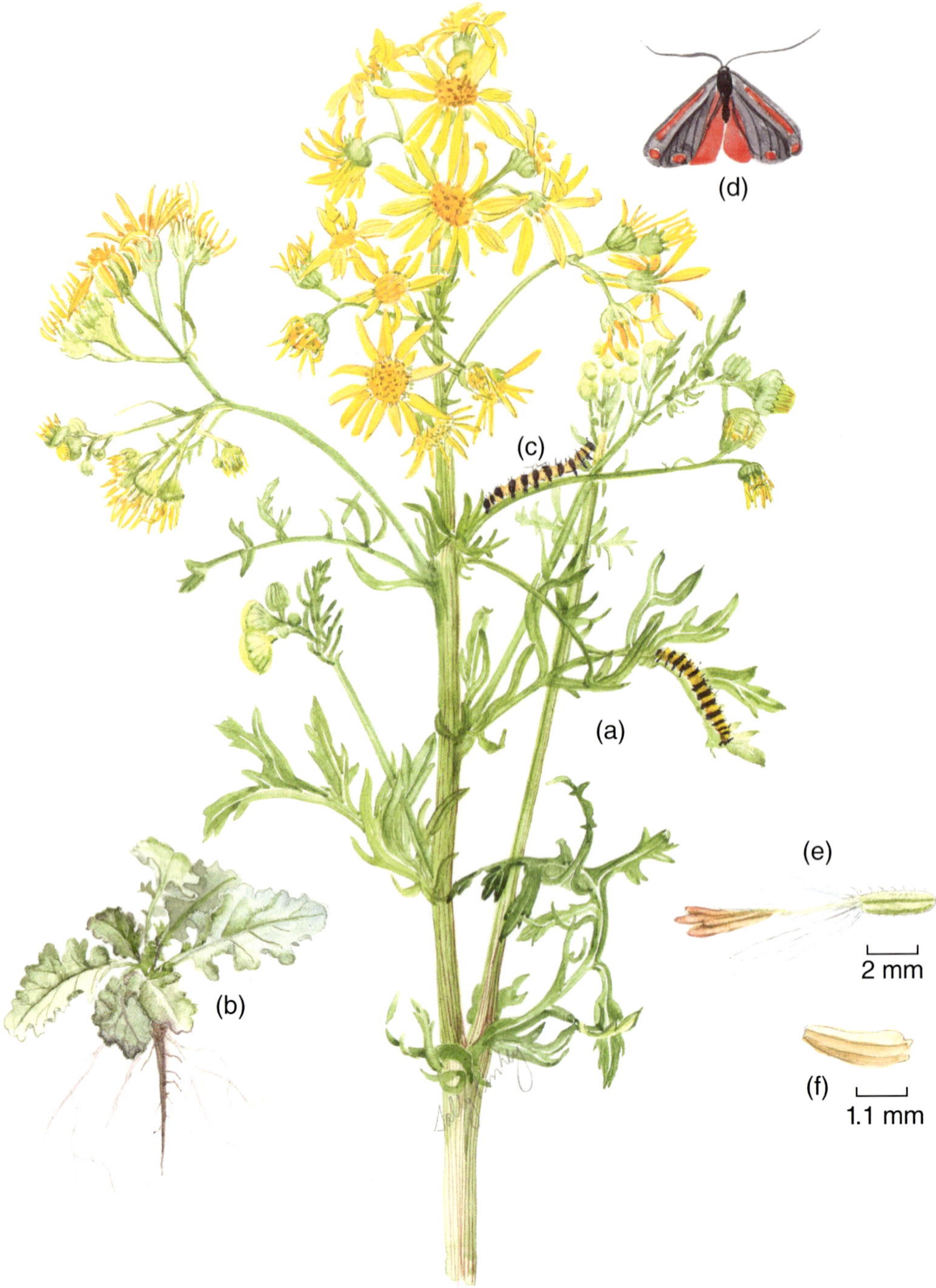

Fig. 7.39. *Senecio jacobaea* plant in flower (a) and seedling (b); cinnabar moth caterpillar (c); cinnabar moth (d); *S. jacobaea* immature seed (e) and mature seed (f).

Senecio jacobaea (Fig. 7.39)

Common name: Ragwort.
Main use: Food for pollinators.
Unsuitability: Invasive, spreading by seed. Toxic in fodder hay.

Environmental benefits

Growing, climate, temperature and rainfall requirements: Robust and ubiquitous plant.
Soil requirements: Likes bare ground.
Soil conditioning: None known, except as ground cover in the event of soil stripped by overgrazing.
Insects: Insect pollinated. Favoured by the cinnabar moth as food for its larvae, which can strip plants completely.
Grazing animals: Normally avoided by grazing animals because of its bitter flavour.

Structure

Roots: Shallow, branched.
Stems: Erect, becoming woody as the seeds mature.
Leaves: Deeply incised, alternate.
Flowers: Yellow corymbs of daisy-like flower heads.
Seeds: Achene 2 mm, ribbed with short hairs. Immature seed has a long white pappus.

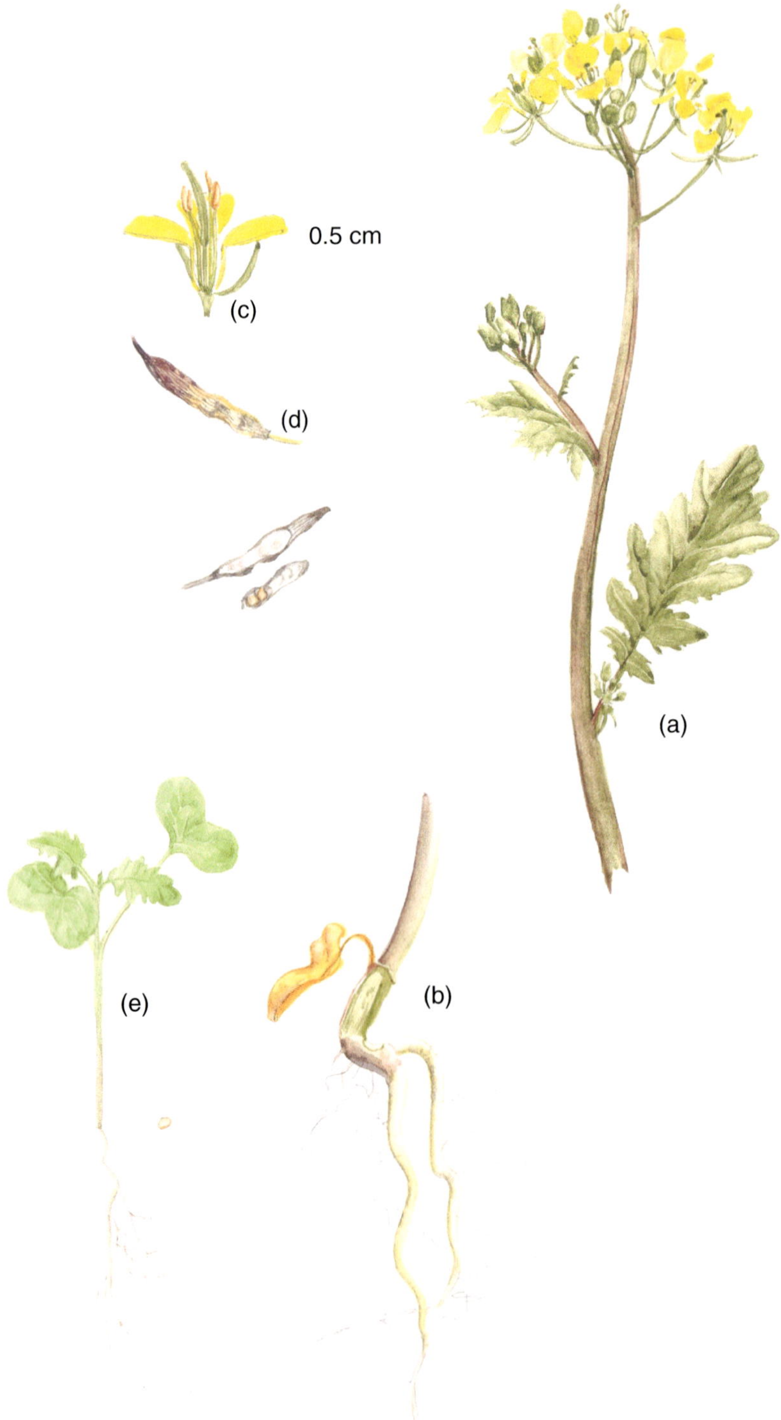

Fig. 7.40. *Sinapis alba* habit (a), root (b), flower enlargement (c), seeds (d) and seedling (e).

Sinapis alba (Fig. 7.40)

Common name: White mustard.
Main uses: Short season cover crop rotation. Quick-growing ground cover for weed suppression. Used for fodder and green manure. The seeds can be harvested.
Unsuitability: Subject to clubroot so should be in the *Brassica* element of the rotation.

Environmental benefits

Growing, climate, temperature and rainfall requirements: Sow in warm spring soil, or in the autumn. Autumn sowings will not set seed. Frost sensitive, and drought and heat resistant. Fast growing (4–6 weeks to flowering).
Soil requirements: Wide range of soil types but particularly likes chalky soils.
Soil conditioning: Minimizes pests in a *Brassica* rotation, suppresses weeds and protects against soil erosion. Reduces pathogens such as wireworm.
Insects: Good resource for early pollinators. Spring sowings can be subject to flea beetle.
Grazing animals: Not normally grazed as pasture but good when cut in a mixed fodder.

Structure

Roots: Shallow, branched.
Stems: Erect, becoming woody as seeds mature.
Leaves: Incised alternate.
Flowers: Yellow corymbs of four-petalled flowers.
Seeds: Round, 2 mm. Light in colour and growing on both sides within the pod.

Fig. 7.41. *Trifolium alexandrinum* habit (a), flower (b), flower enlargement (c) and root (d); *Trifolium repens* habit (e), flower enlargement (f) and root (g).

Trifolium alexandrinum and *Trifolium repens* (Fig. 7.41)

Common names: Berseem or Egyptian clover, and white clover.
Main uses: Berseem clover is a short-term soil fertility booster. White clover may be used long term in pasture, or lower-growing varieties can be used as a permanent green manure mulch between soft fruit rows or vegetable crops. Larger-leaved varieties have better fodder and weed-suppressing qualities. Berseem clover can be used for phytoextraction of several heavy metals.
Unsuitability: Should be mixed with grasses if used for fodder. White clover takes time to establish well.

Environmental benefits

Growing, climate, temperature and rainfall requirements: Sow broadcast or shallow in warm, moist spring soil. May be rolled following sowing to improve contact with moisture. Berseem clover is not frost hardy.
Soil requirements: A wide range of soil types. White clover tolerates waterlogging.
Soil conditioning: These clovers all fix nitrogen well and enrich the soil. A *Rhizobium* innoculant may be recommended to increase the nitrogen fixation of berseem clover.
Insects: Good resource for pollinators and other beneficials.
Grazing animals: Should be in a mixed sward to prevent bloat.

Structure

Roots: Fibrous.
Stems: Soft, not becoming woody.
Leaves: Berseem clover has trifoliate, oblong leaves. White clover has rounder lobes, often with a pale chevron on each lobe.
Flowers: Berseem and white clover flowers are white.
Seeds: Small. Berseem clover has oblong seeds; white clover has triangular seeds.

Fig. 7.42. *Trifolium hybridum* habit (a), root (b), flower enlargement (c), seedling (d) and single flower (e).

Trifolium hybridum (Fig. 7.42)

Common name: Alsike clover (named after the town in Sweden where Linnaeus discovered the plant).
Main uses: Soil improvement, mixed hay and pasture. Attractive to pollinators.
Unsuitability: Needs to be managed as it has the potential to be weedy or invasive.

Environmental benefits

Growing, climate, temperature and rainfall requirements: Sow in spring or autumn. Can stand hard winters and is flood tolerant for up to 6 weeks. Prefers not to be dried out completely in summer.
Soil requirements: A wide range of soil types and well suited to peaty soil with pH 6.0 and soils that are too alkaline for red clover.
Soil conditioning: Nitrogen fixing and useful for sowing beneath cash crops.
Insects: Good resource for pollinators and other beneficials.
Grazing animals: More nutritious in hay than red clover.

Structure

Roots: Shallow, branched tap root with visible nitrogen nodules and wider spread than red clover.
Stems: Weak, smooth and hollow, requiring support from mixed grasses. Grows 30–60 cm tall.
Leaves: Trifoliate, oval and hairless, with toothed margins and stipules.
Flowers: White or pale pink axillary spherical racemes.
Seeds: Short, broad pods 1 cm long containing three to five dull yellowish-green seeds. Cattle spread the seed in dung, but there is little or no vegetative reproduction.

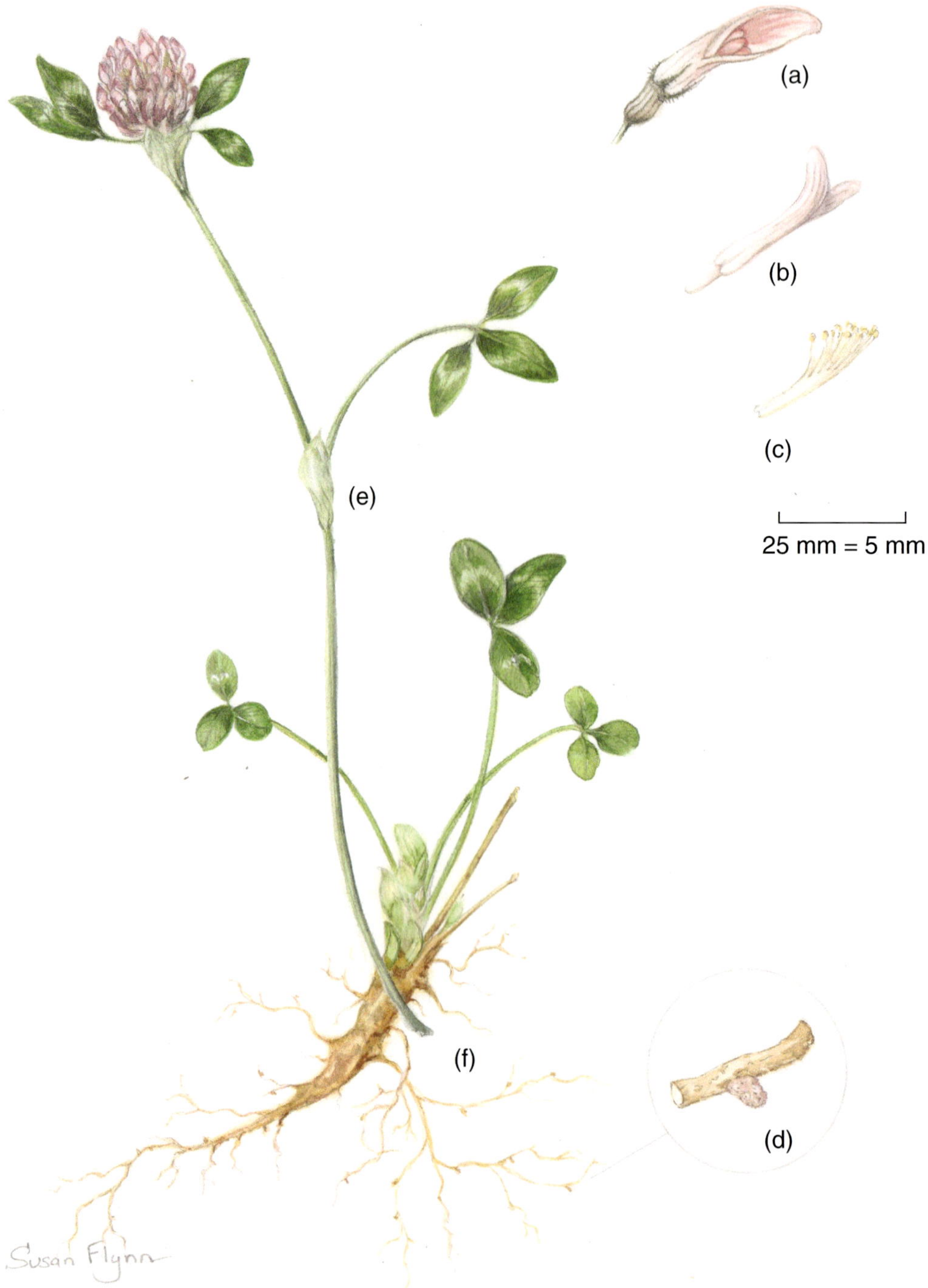

Fig. 7.43. *Trifolium pratense* single flower (a), single flower bud (b), anthers (c), root enlargement (d), flowering plant (e) and root (f).

Trifolium pratense (Fig. 7.43)

Common name: Red clover.
Main uses: Green manure, up to the flowering stage; weed suppression; herbal leys and fodder.
Unsuitability: Grazing exclusively on red clover causes bloating and a reduction in animal fertility.

Environmental benefits

Growing, climate, temperature and rainfall requirements: Sow in shallow drills in spring or late summer if it is not too dry. Largely frost tolerant, surviving for up to 3 years, overwintering as crowns on the roots.
Soil requirements: Wide range of soil types but will need cutting less on a sandy soil. It may need cutting fortnightly on rich soil in a wet summer.
Soil conditioning: When cut or grazed before flowering, it will add substantially to the nitrate levels in the soil. If left in the sward, only about 3% of the manufactured nitrogen returns to the soil. When turned in, it comprises rich biomass, which is increased if mixed with grasses.
Insects: A favourite of pollinators.
Grazing animals: For pasture grazing, it should be mixed with grasses to increase biomass and prevent bloating and loss of animal fertility. The fodder is vitamin and protein rich.

Structure

Roots: Shallow, branched.
Stems: Soft and floppy; do better with support from plant mixtures.
Leaves: Trifoliate.
Flowers: Round pink heads of flowers.
Seeds: Small, 2–3 mm, round or kidney shaped, varying in colour from brown to greenish yellow.

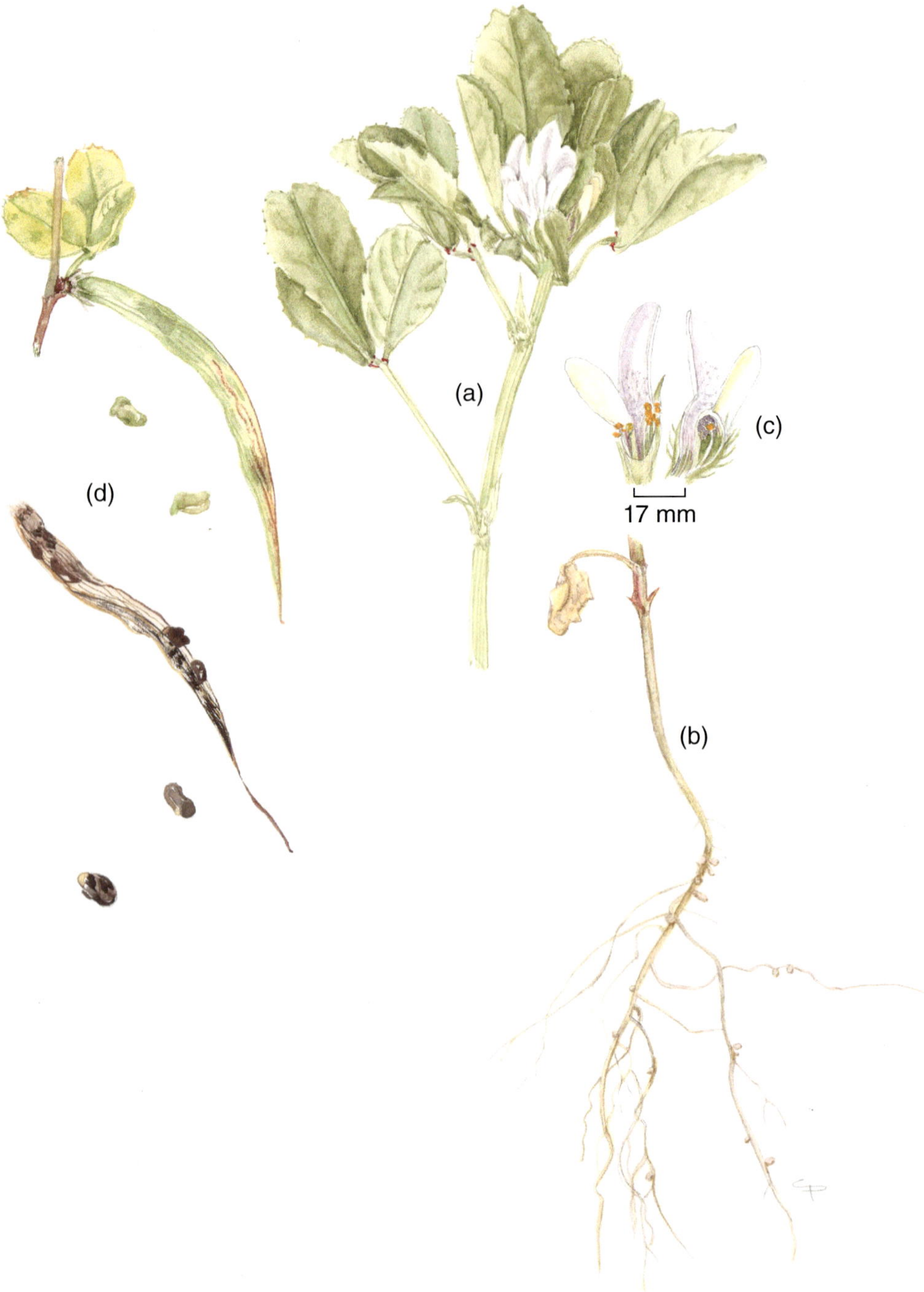

Fig. 7.44. *Trigonella foenum-graecum* habit (a), root (b), flower enlargement (c) and seeds (d).

Trigonella foenum-graecum (Fig. 7.44)

Common name: Fenugreek.
Main uses: Green manure, nitrogen fixer in heavy soil, weed suppressant.
Unsuitability: Toxic in large quantities.

Environmental benefits

Growing, climate, temperature and rainfall requirements: Sow with mixed leguminous green manure seeds into warm spring soil, but in late summer before the soil cools for an overwinter crop. Quick growing. Drought tolerant.
Soil requirements: Well-drained but slightly heavy soil types. Will tolerate lighter ones if fairly moist.
Soil conditioning: Fixes atmospheric nitrogen.
Insects: Good resource for pollinators and other beneficials.
Grazing animals: Should be in a mixed sward to prevent bloat. Containing saponins, it is thought to inhibit the production of ammonia gas in ruminants. There is current research into its use to limit CO_2 emissions.

Structure

Roots: Tap roots and smaller laterals.
Stems: Woody.
Leaves: Trifoliate, oblong, toothed.
Flowers: Racemes may be yellow or pale mauve.
Seeds: Oblong.

Fig. 7.45. *Ulex europaeus* habit (a), flower (b), flower vertical section (c) and seed pod (d); *Ulex gallii* habit (e) and seed pod (f); *Ulex minor* habit (g).

Ulex europaeus, *U. gallii* and *U. minor* (Fig. 7.45)

Common name: Gorse.
Main uses: Game cover, nitrogen fixer and soil stabilizer on wasteland, hedging, good long season nectar source.
Unsuitability: Invasive and tenacious.

Environmental benefits

Growing, climate, temperature and rainfall requirements: Common in temperate climates, but cannot stand repeated hard frosts, so not present at high altitudes.
Soil requirements: Poor soils and light free-draining soils.
Soil conditioning: Fixes atmospheric nitrogen and protects wildlife and saplings.
Insects: Good resource for pollinators and other beneficials.
Grazing animals: Fresh new growth is commonly browsed. Heavy browsing helps to control growth.

Structure

Roots: Deep, branched tap roots and laterals.
Stems: Woody and spiny.
Leaves: Needle-like.
Flowers: Fragrant, yellow legume flowers.
Seeds: Rounded, oblong, in dehiscent pods.

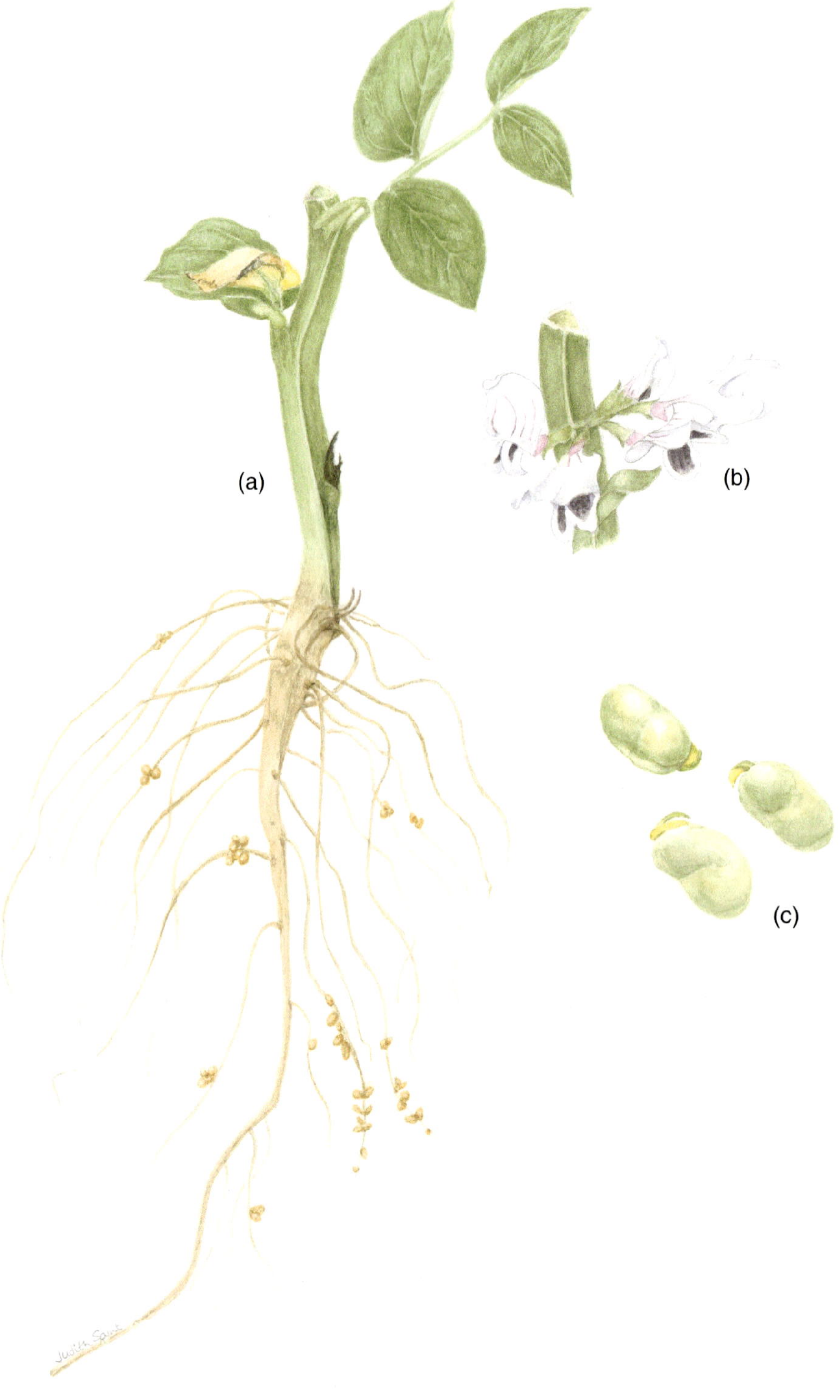

Fig. 7.46. *Vicia faba* habit and roots (a), flowers (b) and seeds (c).

Vicia faba (V. alba) (Fig. 7.46)

Common name: Broad bean or field bean.
Main uses: High in protein. Used for human consumption and for livestock and poultry feed and fodder.

Environmental benefits

Growing, climate, temperature and rainfall requirements: Tolerates acidic soil and short periods of water logging or drought. Prefers cool temperatures. To produce a high yield, early weed removal at 25–75 days after sowing is recommended.
Soil requirements: Well drained, with a neutral pH of 6.5–7.5.
Soil conditioning: Nitrogen fixing with nodules on both the tap and lateral roots. Plant residues incorporated into the soil improve the organic matter content, bulk density and porosity.
Insects: Partially self-pollinating with significant levels of cross-pollination. Honeybees and bumblebees are the main pollinators. Susceptive to aphid and bean leaf weevil.
Grazing animals: Good source of protein for cattle and sheep.

Structure

Roots: Robust, shallow tap root with many fibrous lateral roots reaching up to 90 cm into the soil area.
Stems: Erect, square, coarse, hollow, unbranched, with tiller growth from basal nodes.
Leaves: Alternate, pinnately compound, comprising two to six ovate leaflets up to 8 cm long with toothed stipules at the base.
Flowers: Large, papilionaceous structure with one to six flowers on short auxiliary racemes; white with black/dark purple spots, and fragrant.
Seeds: Long cylindrical pods containing up to six ovoid to oblong seeds with a pale green testa.

Fig. 7.47. *Vicia sativa* habit (a), root (b), flower enlargement vertical section (c), pods (d) and seeds (e).

Vicia sativa (Fig. 7.47)

Common name: Common vetch.
Main uses: Forage, soil improvement.

Environmental benefits

Growing, climate, temperature and rainfall requirements: Sow between the end of March and mid-May for monocropping, but any time in the summer for green manure. When mixed with oats and other grasses, it grows more erect.
Soil requirements: Grassland, farmland, marginal land.
Soil conditioning: Legume and nitrogen fixer. As part of a fodder sward, it is good forage.
Insects: Preferred by bumble bees. Subject to spider mites.
Grazing animals: Good for fattening cattle, and also good for horses if not consumed too abundantly. Animals will prefer vetch to other plants.

Structure

Roots: Fibrous, nitrogen fixing.
Stems: Square and twining. Will sprawl unsupported.
Leaves: Opposite pairs on a compound leaf, ending in a tendril.
Flowers: Single or pairs of purplish pink flowers that grow from the leaf axils in succession.
Seeds: In pods. Diameter 1 mm.
Similar species: Black mustard, brown mustard.

Bibliography

Agricultural Marketing Resource Center (2018) Buckwheat. Available at: www.agmrc.org/commodities-products/grains-oilseeds/buckwheat (accessed 30 September 2021).

Björkman, T., Bellinder, R., Hahn, R. and Shail, J.W. Jr (2008) *Buckwheat Cover Crop Handbook*. Cornell University, Geneva, NY.

CAB International (2021). Available at: www.cabi.org (accessed 30 September 2021).

Cope, T.A. and Gray, A (2009) *Grasses of the British Isles*. BSBI Handbooks, Botanical Society of Britain & Ireland, Durham, UK.

CRC Weed Management (2003) *Equisetum* species. Available at: dpipwe.tas.gov.au/Documents/Horsetails_CRC_wmg.pdf (accessed 30 September 2021).

Karkanis, A., Ntatsi, G., Lepse, L., Fernández, J.A., Vågen, I.M. *et al.* (2018) Faba bean cultivation – revealing novel managing practices for more sustainable and competitive European cropping systems. *Frontiers in Plant Science* 9, 1115.

Rayns, F. and Rosenfeld, A. (2015) *Green Manures – Species Selection*. Garden Organic Factsheet 25/10. Available at: www.organicresearchcentre.com/manage/authincludes/article_uploads/iota/technical-leaflets/green-manures-species-selection.pdf (accessed 27 September 2021).

Rosenfield, A. and Rayns, E. (2018) *Sort Out Your Soil: A Practical Guide to Green Manures.* Cotswold Grass Seeds, Moreton-in-Marsh, UK. Available at: www.gardenorganic.org.uk/sort-out-your-soil (accessed 23 September 2021).

Stoate, C. (2017) The Allerton Project's first 25 years: a rich seam of evidence to support farmland conservation. *British Wildlife* 28, 392–397.

The Wildlife Trusts (2021) Available at: www.wildlifetrusts.org (accessed 30 September 2021).

Tzamaloukas, O., Athanasiadou, S., Kyriazakis, I., Huntley, J.F. and Jackson, F. (2006) The effect of chicory (*Cichorium intybus*) and sulla (*Hedysarum coronarium*) on larval development and mucosal cell responses of growing lambs challenged with *Teladorsagia circumcinta*. *Parasitology* 132, 419–426.

US Department of Agriculture (2021) Plants database. Available at: https://plants.sc.egov.usda.gov/home (accessed 30 September 2021).

8 Soil Stabilizers and Coastal Plants

All plants stabilize soil to some extent, and the greater the diversity of plant cover and microbial activity, the more stable and fertile an area will be. Areas most vulnerable to loss of topsoil are clearly those on slopes. Darwin concluded after careful measurement that worm castings on slopes were seldom incorporated into the topsoil, but were washed away by rain. On dry or degraded land, the first colonizers and protectors of shifting earth are the woody shrubs such as gorse. Shifting coastal sands are often planted with dune grasses with very deep roots that can survive in the environment and limit sand drift inland. Commonly used species are marram grass, sand sedge, lyme grass and sand couch. These are illustrated on the frontispiece and in Fig. 7.4, Chapter 7, this volume).

Almost any plant that can stabilize bare earth, such as the hottentot fig (*Carpobrotus edulis*), which is non-native in northern latitudes, can also be invasive, threatening more delicate native species. Where deliberate courses of action are considered, the balance lies between protecting coastlines and unstable areas, and protecting native species. In a natural course of events, invasive species themselves also alter the soil structure.

There is no doubt that the 74 or so species of mangrove (*Avicennia marina*) play a critical role in coastline protection. While all but one of them are tropical plants, falling outside the scope of this book, which is confined to temperate climate plants, their principal functions deserve a mention. Bill Gates noted that mangroves save 80 billion dollars a year from coastal flood damage. Mangroves break up the eroding power of wave action, provide a safe nursery for marine creatures and collect silt, building up the coastline. *A. marina* grows on the fringes of the temperate climate and as far south as Auckland in New Zealand. There may well be more areas suited to this type of coastal defence as temperatures rise.

Saline Soil

Salt can be leached out of the soil if the soil is deep, permeability is good and there is no water table near the surface, although local wetland and water contamination rules must be met. Over small areas, seedbed preparation and irrigation management can reduce the effects of salinity as deposits concentrate in drier areas between irrigation channels. It is not possible to use heavy equipment on areas with a high water table, but the plants act as a biological pump, lowering the water table and reducing the evaporation that brings the salt to the surface. As the plant residues improve the organic matter in the topsoil, the drainage also improves. Plants have different strategies for coping with salinity. Simply put, these are: avoid, resist or tolerate. It is the plants that have adapted a system to tolerate salt that have a remedial effect.

Food plants bred or adapted to flourish in salty soils will improve it. A collaboration between Dutch farmer Marc van Rijsselberghe and Arjen de Vos from the Free University in Amsterdam produced a conventionally bred potato that is salt tolerant.

DOI: 10.1079/9781789243604.0008

This process has enormous possibilities for coastal land that is increasingly subject to flooding. The plant has adapted to store the excess salt in its leaves, which if then disposed of off site helps to redress soil salinity. The late specialist agronomy researcher Ezra Aberle, working for North Carolina State University, found that barley, rye, sunflowers and sugarbeet (once germinated) could grow in saline soil but did not necessarily reduce its salinity, while salt-tolerant varieties of lucerne did. Plantings were evaluated under both conventional-till and no-till management. Some species of cover crops grown under no-till stabilized soil salinity. Aberle noted that, 'With perennial grasses I've been able to lower the soil salinity. This suggests it could be possible to achieve remediation over time'.

Sea kale (*Crambe maritima*) is salt hardy and can be grown as a crop. Other coastal specialist plants with a market in fashionable restaurants and heritage markets are rock samphire (*Crithmum maritimum*), rocket (*Diplotaxis tenuifolia*), golden samphire (*Inula crithmoides*), buckshorn plantain (*Plantago coronopus*), purslane (*Portulaca oleracea*), sea asparagus (*Salicornia* spp.) and New Zealand spinach (*Tetragonia tetragonioides*).

Purslane deserves a special mention as a superfood containing more omega-3 fatty acid antioxidants than any other known plant. Rich in vitamins A, Bs, C and E, it is used in salads for its acidic flavour or as a spinach substitute. It is less bitter if harvested in the afternoon.

Wheat grasses flourish on saline soil and provide good forage and hay crops. Special varieties of seed cost more, but ordinary varieties also give a positive result. It is mown the first year after sowing at a depth of 3.5 cm, and hay can be cut from it in subsequent years, with the salinity of the soil steadily decreasing.

Sea buckthorn (*Hippophae rhamnoides*) is suitable for land reclamation and stabilizing coastal areas and sand dunes. It is classified in Ireland as a non-native invasive plant but is not prohibited. It is a native of Europe and Asia, with low temperature and salt tolerances, which make it suitable for areas prone to flooding. In a riparian area, it will prevent erosion and improve water quality and ecosystems. It has nitrogen-fixing properties and will improve the quality of the soil, repairing desert areas. It is generally pest free and low maintenance, and makes good hedges and wind screens. The berry is rich in vitamin C, E and B and in fatty acids, which are used in cosmetic and health products. Cold pressing of the berry produces an oil for cosmetics. Because of the thorns, harvesting the berries when they ripen in September is difficult. The easiest way is to cut the whole branch off. Done this way, a commercial crop can only be harvested in alternate years, but it does help to keep the bushes under control. This robust and easy-going plant earns its keep in marginally cultivable habitats.

Steep mountainous slopes and retaining embankments can be stabilized by plants with a network of strong running roots. Field horsetail (*Equisetum arvense*) does the job on poor and waterlogged soils but is also very invasive. Large bamboo species hold whole Asian mountainsides together, while also providing a crop for building materials. Of the 1400 species of bamboo, the most suitable for a temperate climate is Scottish bamboo (*Phyllostachys humilis*). It is frost hardy and will grow in shade, and on clay or sandy soil, and can withstand short droughts once established. All plants with a strong root system that establish quickly and need little care have the potential to be invasive and this must be weighed up with other considerations and strategies. New embankments in danger of erosion can be covered with a biodegradable mesh and sown with a permanent vegetation cover. The mesh protects the embankment while the plants are establishing.

Fig. 8.1. Example of a badly eroded arable field.

Loss of topsoil as a result of seasonally bare-earth farming practices (Fig. 8.1) is mitigated by planting winter cover. Plants suitable for this also fall into the category of green manures and nitrogen fixers (see Chapters 2 and 3, this volume).

Other areas subject to erosion by water are river banks, where alders, willows and poplar roots, in addition to sustaining vigorous growth, create an organic retaining network, which is also a specific and species-rich habitat. Together with birches and aspen, these early colonizers of disturbed or degraded ground stabilize soil carbon and create an organic topsoil.

Grasses for Soil Erosion

Native grass plants are useful for erosion control and have the added benefit of fitting readily into the landscape. They will easily transplant into conditions that mimic their natural habitat. Native grasses also need less maintenance as they are adapted to the region in which they occur and receive most of their needs from the existing site. The right grasses for soil erosion depend on the zone and region. Overall, some excellent choices are: timothy (*Phleum pratense*), foxtail (*Alopecurus pratensis*) and smooth brome (*Bromus racemosus*).

Fertile Islands

Fertile islands, which are patches of land that are left under a diverse shrub-dominated plant cover, helps arid grazing land to maintain its fertility. Shrub soils tend to be more porous and contain more organic carbon, nitrogen and phosphorous than the grass area, supporting a robust microbial community. The best shrubs are those with leaves that decompose quickly. Greater amounts of compostable litter, together with higher soil pH, are known to enhance the presence of a well-developed network of fungal hyphae, maximizing nutrient cycling and sequestration rates. The effect of balanced grazing accentuates the natural development of shrubby fertile islands, but severe overgrazing stresses the soil microbial communities to the point at which they can no longer buffer the effects of overgrazing. The inevitable result is bare soil, land degradation and loss of soil to erosion. The interaction of plant and animal behaviours point to an optimum management strategy of little interference in natural processes in dry grazing lands, provided the burden of grazers is not too great.

Recent studies demonstrate that increasing aridity reduces the diversity and abundance of soil fungal and bacterial communities, damaging key ecosystems essential to nutrient cycling and plant production. However, the extent of this damage in terms of recovery prospects has yet to be evaluated.

Bibliography

Cai, Y., Yan, Y., Xu, D. Xu, C., Wang, C. *et al.* (2020) The fertile island effect collapses under extreme overgrazing: evidence from a shrub-encroached grassland. *Plant and Soil* 448, 201–212.

Darwin, C. (1881) *The Formation of Vegetable Mould Through the Action of Worms.* John Murray, London.

Egan, G., Crawley, M.J. and Fornara, D.A. (2018) Effects of long-term grassland management on the carbon and nitrogen pools of different soil aggregate fractions. *Science of The Total Environment*, 613–614, 810–819.

Gates, B. (2021) *How to Avoid a Climate Disaster*. Allen Lane, London.

Nickel, R. (2017) New life for saline soil. *Successful Farming*, online. Available at: www.agriculture.com/crops/cover-crops/new-life-for-saline-soil (accessed 29 September 2021).

Ochoa-Hueso, R., Eldridge, D.J., Delgado-Baquerizo, M., Soliveres, S., Bowker, M.A. *et al.* (2017) Soil fungal abundance and plant functional traits drive fertile island formation in global drylands. *Journal of Ecology* 106, 242–253.

Spalding, M., Kainuma, M. and Collins, L. (2010) *The World Atlas of Mangroves*. Earthscan, Abingdon, UK.

Zaninovich, S.C., Montti, L.F., Alvarez, M.F. and Gatti, M.G. Replacing trees by bamboos: changes from canopy to soil organic carbon storage. *Forest Ecology and Management* 400, 208–217.

9 Weeds and Invasive Plants

> The principle methods of non-chemical control for perennial weeds are cultivation, cutting or grazing and competition. The aim is to prevent seeding and to starve out the perennial organs.
>
> Defra project OF0367 (2008), *A Review of the Management of Selected Perennial Weeds*.

It has often been said that a weed is a plant where you don't want it. Some plants are more of a nuisance than others in the wrong place. The same is true of invasive species, and while they sometimes prove more troublesome than weeds, it should be acknowledged that the terms 'weed' and 'invasive species' are labels given to them by people for control purposes. These are the plants that are the opportunists, colonizers and survivors. In the rough and tumble world of plant survival, some have a value, even if they are inedible. Some will cover bare ground quickly and reduce erosion. Some plants considered garden weeds such as ribwort plantain (*Plantago lanceolata*) and bird's-foot trefoil (*Lotus corniculatus*) are valuable components of a herbal lay. Deep-rooted plantain is a drought survivor and valuable forage, while bird's-foot trefoil will smother other weeds and enrich the soil with nitrogen at the same time. Traditionally, any volunteer plant growing in a crop is thought of as a weed and a crop competitor. While this is true of plants that germinate at the same time and grow at the same rate, using the same nutrients, it is not necessarily true of all 'weeds'. Not all plants have the same requirements and many can coexist without competing. They may enhance each other. This is known as the 'Darwin effect'. By careful experimentation, Darwin proved beyond doubt that mixed crops had 50% better yields. Where chicory (*Cichorium intybus*), ryegrass (*Lolium* spp.) and Dutch or white clover (*Trifolium repens*) were grown together, the yield was not the average of each but much more, because of their differing habits in terms of root depth, nutrition and moisture needs.

When tackling infestations of weeds, it is important to know how they propagate, whether vegetatively or by seed. Soil disturbance can be minimized for those weeds that multiply by root segmentation and become an unwelcome infestation. The Weeds Act of 1959 requires landowners to prevent the spread of couch grass, broad-leaved and curled dock, and creeping thistle.

Some plants thought of as invasive, such as gorse and bracken, are the first to colonize newly exposed land, protect coasts and begin a basic ecosystem. These are often the plants that need little nutrition and are tolerant of toxins. Gorse (*Ulex europaeus*) is itself a legume and will build up nitrogen in the soil. Left to themselves, such plants will protect wildlife and saplings, which will eventually grow and shade out the first rough ground cover. Thus, while this long process is unsatisfactory on the small time and space scale of annual production, in the larger scheme of things, it is not a problem. Invasive species become a danger where they are newly arrived in a habitat where they have no natural predators. Species that have developed robust growing and survival techniques against predators in their native environment quickly dominate less robust native species in the absence of predators in their adopted environment, such as Himalayan balsam (*Impatiens glandulifera*). Where vigorous destruction methods fail, it falls to scientists to evaluate possible dangers in introducing the natural predators of the invasive species to its new environment.

DOI: 10.1079/9781789243604.0009

The following sections describe some of the more common weeds and invasive plants and methods to control them.

Black-grass (*Alopecurus myosuroides*)

This is a widespread arable weed that has become resistant to herbicides such as Roundup. It grows in damp, heavy soils, and can quickly infest a winter wheat crop. It can be a real problem for farmers when it contaminates the crop. It usually germinates in moist conditions in September/October, the same time as winter wheat. Control methods include the following:

- Improved drainage and ploughing once every 5 or 6 years help to control it, but the prevalence of autumn-sown crops in the UK is the main reason why black-grass has become an increasing problem, as it germinates at about the same time.
- Matt Broadhurst, who farms in Bedfordshire, UK, controls his black-grass with a late spring sowing of soybeans. This low-input crop, which needs no fertilizer, gives time to control the black-grass while also improving the soil.
- Leaving fields fallow for a season will reduce the severity of the infestation, but better still, a 2-year ryegrass (*Lolium* spp.) and red clover (*Trifolium pratense*) green manure will suppress germination of the black-grass, and reduce its population. Ryegrass can also be a cash crop for quality hay or silage if used in this way on an arable farm.

Bracken (*Pteridium aquilinum*)

This prolific plant has benefited from woodland clearance and reduced upland management to the extent that it is a potential nuisance on most well-drained open ground, uplands and marginal land. Spreading mainly by robust underground rhizomes, a single plant can cover a whole hillside. It is toxic to cattle and to some extent sheep, although it will be avoided if there is sufficient alternative grazing. As the spores are also toxic, reduced use of dried fronds as animal bedding is probably an improvement.

Bracken provides dense cover for some wildlife, but harbours ticks, which may carry Lyme disease.

Chemical control is no longer approved (see Appendix), but many traditional control methods are effective:

- Bruising the new fronds in spring with a bracken basher is a recognized procedure, as is cutting at 6-week intervals between June and September.
- Any damage inflicted on the bracken in the early growing season will reduce its strength. Thus, ploughing, rolling, harrowing or trampling by sheep, cattle or horses while the fronds are young reduces growth, while overgrazing encourages it by reducing the competition of palatable grasses. Bracken will not survive heavy frosts, wind damage or salt spray. Free-range wild pigs will also root it up quite effectively.
- A strong ground cover that can compete with the bracken will hinder its establishment and at the same time provide rough pasture. Species that can do this are creeping soft grass (*Holcus mollis*) and cock's foot (*Dactylis glomerata*).
- Complete clearance using a combination of these approaches takes about 7 years, but the bracken will quickly re-establish if the treatments are discontinued before it is eradicated.

Common couch (*Elymus repens*, syn. *Elytrigia repens*)

Couch is at home in both light and heavy soils but spreads more readily in tilled light soils that are low in organic content but high in nitrogen. It is only a problem in cereal crops, peas and oilseed rape, where it reduces the crop yield. Infestation by seed can occur in poor-quality seed sources, and can be exacerbated by the action of a combine harvester. The seed is viable in the digestive systems of cows, sheep and horses, and will germinate in the droppings of ruminants. The main danger is that it hosts many pathogens, such as frit and gout flies, and fungal diseases such as ergot, barley leaf blotch, cereal wilt and take-all root rot. Barley may be attacked by a root lesion nematode hosted by couch grass rhizomes. Any form of biological control is likely to damage the main crop as well, and increasing nitrogen application to counter fungal damage is counterproductive.

- One defence is long and varied rotations cultivating a phosphate-rich soil with an abundance of microbe life. Defra in the UK suggests mechanical removal after harvest. Rotovation will bring rhizomes to the surface to dry out. They can then be raked up and burnt. Residual rhizomes will shoot again, so further rotovations at 3-week intervals until winter begins are effective treatment. The treatment needs to be repeated the following year. This approach precludes the opportunity of an autumn sowing of ground cover that would discourage the weed.
- Couch can be shaded out, and a smother crop of lucerne or clover could be the least costly and laborious way of controlling it.

Couch grass is palatable and does not thrive in grazed pastures. Pigs, horses and cattle will also seek out the rhizomes to eat. It is seldom a nuisance in permanent pasture.

Creeping thistle (*Cirsium arvense*)

Creeping thistle flourishes where there are high levels of nitrogen, becoming a problem in arable crops where it reduces yields in all but spring cereals. It is generally present in grassland, and has become more so in 'set-aside' fields. Some seed is viable in cow manure and bird droppings, and may be present in uncertified arable seed, and spread by a combine harvester, but most regeneration comes from a single root stock. The plant uses the roots as winter food storage, easily regrowing from cut stems and readily spreading from adventitious roots. Creeping and other thistles host bean aphids, mangold fly, celery fly and swift moth larvae, together with mildews affecting lettuce and chrysanthemums. Many birds, especially finches, feed on the seeds. There are indications that creeping thistle is allelopathic particularly towards sugarbeet and wheat, inhibiting growth. The thistle is particularly aggressive in a nitrogen-treated barley crop. The brittle roots are easily broken and spread by cultivation, exacerbating the problem. Close grazing by sheep and rabbits reduces the competition from other plants and allows creeping thistle to spread, while cattle grazing leaves the sward longer and offers fewer opportunities. Goats, donkeys and llamas will also eat the thistles and help to control them.

Traditional methods of control have been to destroy the tops to starve the roots:

- Regular hay cutting prevents seedings and eventually weakens the plant. Cutting from July onwards is most effective, especially in a fertility-building legume crop, which smothers emerging thistle shoots.

- Mechanical pulling will also eventually weaken the infestation, but the effect only lasts for a few years.

Investigations into biological controls involving flea beetles, a stem-mining weevil and a gall fly have been disappointing, but the painted lady butterfly larvae can defoliate stands of creeping thistle in a good migration year. Butterflies are easy to raise from caterpillars as their nests are clearly visible on the thistle plants. The thistle is also the food plant for some species of micro-moth larvae. Rust fungus can eliminate colonies of the thistle following attacks by the stem-mining weevil, which weakens the plants initially. The rust fungus *Puccinia obtegens*, which occurs worldwide, weakens and stunts the thistle, and may eliminate it. Creeping thistles inoculated with *Puccinia punctiformis* rust fungus usually die before flowering, and a summer treatment on young plants can establish a systemic infection.

The wide range of pathogens and wildlife that feed on creeping thistle offers opportunity for research into one that will control the thistle without affecting crops. It also makes the obvious point that a rich mix of soil microbes and insects will reach a balance without extra management if they are not destroyed by agrichemicals. Meanwhile, light grazing on a diverse sward is a route towards that balance.

Docks (*Rumex obtusifolius*, *R. crispus* and the hybrid *R.* × *pratensis*)

There are two main species of dock: broad-leaved (*Rumex obtusifolius*) and curled (*R. crispus*). They tend to hybridize, creating great variety, but as far as infestations are concerned, there is little difference in nuisance or treatment. There are few soil types or climate extremes that dock cannot withstand, and it has strategies to cope with waterlogging, pollution, trampling and mowing. Secondary taproots can form a large area of cloned plants where seedlings are discouraged. It seeds freely and will germinate at variable times in any bare patches caused by overapplications of slurry, flooding or ploughing, but seedlings cannot establish well in dense grassland. Seed is often found in home-saved seed but not in merchant's seed. Seed viability is maintained during cattle digestion and the high temperatures of silage and slurry. Docks tend to be a problem in the early phases of conversion to organic farming, but their density decreases after a few years. Regular cutting has little effect on mature plants, which regrow readily from their crowns. The plants host the potato tuber eelworm, bean aphids and mangold fly.

Docks are a valuable component of mixed grazing where it is not too dense to inhibit grass growth. The inclusion of dock in a substantially legume forage will prevent bloat.

- Grazing by sheep and goats reduces docks, but cattle grazing does not.
- Hand removal with a docking iron is time consuming but is possibly the most effective method of reduction. It is easier when the soil is damp and when the seeds are beginning to swell and the root is shrinking. Most roots cut at a depth of 8 cm will not regenerate. Uprooted docks should not be left where they can grow. They can be burnt or dried out for 4–8 weeks.
- Where biological control is attempted, it is best to select a native species of predator and augment it. Native stem-boring weevils, larvae of the leaf-mining fly and the chrysomelid beetle are possible candidates. Left to multiply, the beetle can significantly

reduce the dock infestation, but its population will be adversely affected by field applications of pesticide, and by cutting and mowing the docks.
- Two varieties of rust fungus affect the dock without reducing it much. The presence of rust fungus inhibits an increase in chrysomelid beetles. Honey fungus will kill dock, but it should not be introduced deliberately.
- Undersowing crops with clover discourages seedling establishment, as does maintaining the density of permanent pasture.

No control is fully effective on its own, but natural controls at a background level help with management. Encouraging all of the natural control agents is likely to prove beneficial.

Field horsetail (*Equisetum arvense*)

Field horsetail will grow on most soil types, and while not a particular problem on arable land, it is troublesome in fruit and perennial crops. Its roots penetrate to 1.5 m, and it accumulates heavy metals such as cadmium, copper, lead, zinc and gold in its tissues. It can be used to make fungicide for treating blackspot and mildews, and the roughness of the silica in it makes it a good saucepan scourer. It is toxic, both fresh and dried, to sheep, cows and horses, and is therefore a danger in pasture.

As it is a living fossil, with top survival credentials, it is more like a fern than an angiosperm, and has a life cycle similar to a fern, producing spores from a cone. It multiplies mostly vegetatively from sections of rhizome, which generate if cut by field disturbance. It is encouraged by soil compaction and long periods of arable rotation although without affecting arable yields to any great extent, as it does not compete for nitrogen.

- Hoeing the young shoots before the spores shed helps to control its spread and weakens the plant.
- Horsetail can be shaded out by a dense crop, so can be controlled by early crop sowing ahead of its development, as well as by undersowing.

Gorses (*Ulex europaeus, U. gallii and U. minor*)

The Old English 'gorst', means a wasteland or uncultivated area. As a member of the Fabaceae, gorse fixes nitrogen nodules on its roots, enabling it to live in areas of poor soil quality. However, this can acidify the soil, making it difficult for other plants to grow. It is common to disturbed areas, grasslands, forest margins, coastal habitats and wasteland. There are three species in Britain. Common gorse (*Ulex europaeus*), flowering in spring, is the most familiar, widespread and invasive, being present on every continent. It has the most robust growth character. Western gorse (*Ulex gallii*), flowering in autumn, is frequent on the western coasts of Europe. It is relatively low growing, yet robust. Elsewhere, dwarf gorse (*Ulex minor*) is a low-growing, sprawling shrub that is an occasional component of European dry heath. It is commonly grazed.

Gorse is classed as an invasive species and lives for up to 25 years. It is tenacious once established, and can outcompete and displace other native plants. It is also a fire hazard as it holds on to its dead and dry branches and leaves. Gorse scrub occurs wherever soils are light and free draining, in areas that are relatively free from severe frosts. While it is very important for wildlife, it can encroach on otherwise cultivable land. Old and degenerate gorse is relatively poor for wildlife. However, the accumulation of plant debris increases soil fertility, encouraging bracken. The accumulated dead material also presents an increased fire risk. Very old, leggy gorse rarely regenerates when cut.

The flowers can be used to produce a yellow dye, or can be eaten, and compact hedges provide an important refuge for endangered wildlife in winter. The long flowering period also makes it an important nectar source. Gorse hedges can be maintained by regular trimming, but old and degenerate gorse is no use to wildlife.

A large bank of seeds usually survives in the soil, so clearing the gorse and removing the loose organic litter allows the seeds to germinate. Burning *in situ* can also be dangerous, because of the high volume of very combustible material. The fire will expose the seeds and heat them, encouraging germination.

Gorse can be restricted or removed relatively easily using a number of techniques, depending on local conditions:

- Cutting to ground level and letting livestock, deer or rabbits browse the regeneration is often effective where the surrounding vegetation has low palatability. High numbers of livestock can, however, compromise sensitive vegetation and vegetation structure.
- Repeated cutting will eventually kill gorse but may take several years and so be expensive.
- Grubbing out whole bushes with the rootstock is effective but can create conditions for gorse to recolonize.
- Small patches and individual bushes are usually best cut by chainsaw or clearing saw, but it is more economical to flail large stands, although removing the large volume of shredded gorse is likely to be a problem unless a cut-and-collect machine is used. Dead gorse and litter can be burnt off site to avoid regrowth from seed. Most cut stumps will regenerate within a year unless grazed and resown with an alternative robust plant to outcompete. Bracken should be controlled at this stage.

Japanese knotweed (*Reynoutria japonica*)

This invasive plant is a member of the Polygonaceae, the same family as buckwheat (*Fagopyrum esculentum*). It is tolerant of a wide range of soil types with strong rhizomes up to 3 m deep, making it resilient to eradication.

- Digging out is rarely completely successful, and disposal of the roots, classified in the UK as 'controlled waste', can be expensive.
- Regular cutting weakens the growth, as does glyphosate poured into the hollow stems.
- In Japan, the principle predator is the psylid insect *Aphalara itodori*, which feeds exclusively on Japanese knotweed. Early trials of the introduction of this species to other countries show promise.
- Goat grazing and pig rooting also help in the control.

The flower is an important source of nectar for honey bees in late summer and early autumn when there are fewer flowers around. Birds also eat the seeds but are not known to disperse them in a viable state. Young stems are edible as a vegetable after soaking, peeling and boiling. It is used in herbal medicine for its compound resveratrol, which is an antibiotic and has recently been found by the Johns Hopkins Bloomberg School of Public Health to be effective in treating Lyme disease where other antibiotics have been unsuccessful.

Ragwort (*Senecio jacobaea*)

This native British plant has had an unfortunate press. It is true that if it gets into hay forage in large quantities it is toxic, but on the whole, it does no more damage in pasture than any other poisonous wild plant because it tastes bitter and grazing animals avoid it. It is not true that British farmers are legally obliged to eradicate it. The UK Weeds Act of 1959 and the UK Ragwort Control Act of 2003 merely provide for orders to be made in the case of a nuisance, but in the absence of an order, there is no compulsion. The plant is a favourite of pollinators and greatly enhances the biodiversity of the habitat. The larvae of the cinnabar moth in particular are voracious feeders, sometimes eating the whole host plant, absorbing the bitter taste of the plant as a defence against predators. The moth, which is black and red and flies during the day, has been successfully introduced to New Zealand as a biocontrol and in the USA in conjunction with the ragwort flea beetle.

- To reduce the incidence of ragwort on land, pulling up plants before they set seed will prevent it getting into hay or seeding, but as the seed will only germinate on bare ground, the best way to reduce it is to keep the ground covered with pasture.
- Overgrazing is a great encouragement to ragwort and should be discouraged.

Pulling the ragwort is an opportunity to cover the ground and prevent reseeding by dropping a herb mixture of seed and compost into the hole and treading it down. A mixture of chicory (*Cichorium intybus*), salad burnet (*Poterium sanguisorba* subsp. *sanguisorba*), sheep's parsley (*Petroselinum crispum*) and ribwort plantain (*Plantago lanceolata*) will greatly improve the variety and medicinal qualities of the sward, especially for horses.

Rushes

There is concern about the increased areas of grazing land covered by rushes in the northern hemisphere (*Juncus conglomeratus* and *Juncus effusus*), as indicated by Google Earth between 2005 and 2018. The problem is that the rushes are not edible to livestock, constitute a fire risk and spread vegetatively as well as by prolific seed with high viability. Hard rush (*Juncus inflexus*) poisons sheep and cattle, causing blindness and death in those that acquire a taste for it. The reason for the spread of rushes is unproven, but suggestions are that ground that is overgrazed, laid bare by ploughing or impacted by heavy machinery offers opportunity to invasive rushes.

- While as yet no wholly satisfactory method of control has been developed, targeted improvement of the soil structure will discourage growth. Cutting before flowering in May also helps.

Bibliography

Bond, W., Davies, G. and Turner, R.J. (2007) The biology and non-chemical control of common couch (*Elytrigia repens* (L.) Nevski). Available at: www.gardenorganic.org.uk/sites/www.gardenorganic.org.uk/files/organic-weeds/elytrigia-repens.pdf (accessed 15 October 2021).

CABI summary data. Available at: www.cabi.org/ (accessed 14 October 2021).

Clapham, A.R., Tutin, T.G. and Warburg, E.F. (1962) *Flora of the British Isles*. Cambridge University Press, Cambridge, p. 332.

Darwin, C. (1881) *The Formation of Vegetable Mould through the Action of Worms*. John Murray, London.

Defra (2003) Effects of grazing management on creeping thistle and other injurious weeds and integration of grazing with weed control – BD1437. Defra, London. Available at: http://sciencesearch.defra.gov.uk/Default.aspx?Menu=Menu&Module=More&Location=None&Completed=0&ProjectID=9115 (accessed 29 September 2021).

Defra (2007) *Code of Practice on How to Prevent the Spread of Ragwort*. Defra, London. Available at: www.gov.uk/government/publications/code-of-practice-on-how-to-prevent-the-spread-of-ragwort (accessed 23 September 2021).

Le Duc, M., Pakeman, G., Putwain, P. and Marrs, R. (1999) The variable responses of bracken fronds to control treatments in Great Britain. Available at: http://randd.defra.gov.uk/Document.aspx?Document=BD1226_1877_FRA.pdf (accessed 29 September 2021).

Morse, R. and Palmer, R. (1925) *British Weeds and their Identification and Control*. Ernest Benn, London.

Pakeman, R.J. and Marrs, R.H. (1993). Bracken. *Biologist* 40, 105–109.

Price, R. (2019) Expert tips to control soft rush in grassland. *Farmers Weekly,* online. Available at: www.fwi.co.uk/livestock/grassland-management/expert-tips-to-control-soft-rush (accessed 11 November 2020).

Streeter, D., Hart-Davies, C., Hardcastle, A., Cole, F. and Harper, L. (2009) *Collins Flower Guide*. HarperCollins, London.

Wandeler, H. and Bacher, S. (2006) Insect-transmitted urediniospores of the rust *Puccinia punctiformis* cause systemic infections in established *Cirsium arvense* plants. *Phytopathology* 96, 813–818.

10 Hedges and Trees

> Agroforestry is about making the most of land area through three dimensional farming.
>
> Stephen Briggs (2012).

Traditionally, hedges that define fields, provide windbreaks, constitute barriers to livestock and provide habitat for wildlife are a distinctive feature of an agricultural landscape.

However, in the UK alone, some 250,000 miles of hedges were uprooted in the second half of the 20th century, comprising about a third of the total. Similar changes took place in most temperate zones following the trend to larger-scale farming with large machinery. Soil erosion on an epic scale followed. A hedge prevents wind soil erosion, protecting up to 20 times its height downwind and up to 30 at ground level. A hedge at its natural height of about 12 feet will maintain itself and protect wildlife, but overcut square flailed hedges are no good. They can sustain nothing, not even themselves. The collateral damage is cumulative. Even the decline in bats has been accelerated by loss of hedges. An animal that navigates by echolocation is lost in an open space. Hedges and trees are its highways.

The border to a hedge is also significant. It adds to the habitat of the hedge, providing protection and food for its residents. A dual purpose strip of, say, 6–10 m next to the hedge can be sown and harvested as hay while providing a turning area for tractor passes for the main crop. Because most of the compaction occurs when the machinery turns, this area is less productive for the main crop so there is little productivity loss. Field margins also allow hedges and trees to root into the permanent lay, and can be a riparian zone between water courses and crops. UK definitions for their stewardship scheme are 6 m for a buffer strip, 6–10 m for a set aside strip and 6–25 m for conservation headlands.

There is no reason why hedge produce should not be harvested if it is marketable. Damsons and the Myrobalan red (cherry plum) do well in hedges. Although the Myrobalan red may fruit erratically, it requires little maintenance and is easy to harvest. These fruits, which are traditionally used for preserves, have not recently fitted in to the wholesale nature of food production and marketing, but the time is now right for a return to more local smaller-scale food production.

Hazel (*Corylus avellana*) and Filbert (*C. maxima*) nut trees make excellent hedges, doing well when trimmed to hedge height and also producing a useful crop. Cultivated since pre-history right across the northern hemisphere on poor well-drained soil, there are hundreds of varieties of improved hazel/cob and filbert nuts suited to commercial cultivation. They give a profitable yield after 2 years, and where the soil pH is right, truffles can be grown in association. Indigenous and naturalized hazels can be grown from seed or rootstock. Layering is also an option, and special varieties can be grafted. They are not self-fertile, so a hedge should contain plants from varied stock. For minimum maintenance, take stock from local varieties and check marketing opportunities before selecting a variety, as there is a lot of difference in size, shell thickness and the ease with which the husk separates from the nut. Harvesting can take place from late summer onwards.

DOI: 10.1079/9781789243604.0010

Sea buckthorn (*Hippophae rhamnoides*) is a robust and salt-resistant shrub that produces a marketable superfood with high nutrition and vitamin qualities, while providing a dense hedge. Like a gorse hedge, it does need some containment (see Chapter 8, this volume).

Timber can also be a crop in a hedge. While some hedging machines are not adapted to interruption by a mature trunk, long-term investment in carefully spaced and trimmed oaks, for instance, provides forage and constitutes capital for both a family and a nation.

Walnut and sweet chestnut trees also flourish in a hedge line and produce a valuable crop. Sweet chestnut grows best in slightly acidic soil. Traditional farming practices in France, where land ownership persisted through generations, make a distinctive landscape of hedges with straight mature trees at intervals, causing minimal shading and adding greatly to the value of the land. Both historically and today, France exports sustainable timber to countries with a shorter-term approach to hedge diversity.

If hedges are considered part of the diversity plan on a farm, providing free fencing, windbreaks and a fertile island, and protection for wildlife as well as a crop, then replacement of previously uprooted hedges seems an obvious win–win project.

For farmers with short tenancies, fruit trees in a north/south configuration to prevent shading give a quicker return without shading the main crop. An 8% coverage of trees still leaves plenty of space for a main crop, while in time exceeding the total output. Trees use light at different times of the year to cereal crops, as fruit ripens after the main harvest, and moisture for trees and hedges is drawn from well below the soil depth that supplies the cereal crop. Early circular cutting around the lateral roots encourages the roots to grow down. Strips of arable or vegetables two tractors' widths between each row of trees keeps the sown areas easy to maintain, while a mower between the trees controls weeds and encourages a succession of pollinator plants. Coppice trees like hazel/filbert and willow will do well in this strip system and supply a high biomass. There are many trees that provide a nutritious food crop, which is the future for combined land restoration with food provision for an increasing population. The choice of trees should be based on varieties of indigenous species. Mixed planting reduces the incidence of pests and the necessity to take pest control measures. Low expenses means that harvests are almost pure profit.

It is known that livestock does better on a mixed pasture that supplies a variety of nutrition. Livestock will also graze hedges where they can, and grazers will show a preference for certain species depending on their needs, and in so doing self-medicate to keep healthy.

Silvopasture is the name given to the practice of grazing livestock in woodland and is favoured for its natural approach to organic food production. For this, the trees need to be mature or protected, but it is low maintenance and scores high for long-term sustainability. Grazing and trampling in mixed or light woodland controls invasive species such as bracken and gorse, allows light into clearings and encourages grass. Poultry do well in woodland and are suitable for a young plantation.

Both hedges and tree canopies create an area of greater fertility caused by organic litter and biochemical nutrient recycling. Alders also fix nitrogen and do nearly as well away from river banks as on them. The trees provide increased infiltration rates which reduce surface runoff, water erosion and loss of nutrients suspended in stream water. Thus, there are several biogeochemical processes by which hedges and oak trees concentrate nutrients and maintain islands of enhanced fertility beneath and beyond their canopies. The reservoir of associated soil organisms is available to gradually recolonize adjacent areas depleted by large-scale monocrop farming.

Since the dawn of farming, it has been known that sloping land is unsuitable for ploughing and, being more subject to soil erosion, should be kept covered in trees. Indeed, the loss of land by landslip in the Himalayas may be so immediate following tree removal that cause and effect are catastrophically evident. In some countries, traditional farming methods accommodate the necessity to conserve tree covering on slopes, and there are many examples of successful agricultural systems that achieve this. Pigs are experts at fattening themselves on tree produce not favoured as human food, such as wild cherries, horse chestnuts, acorns and holly leaves. They disturb the soil and bring a whole new range of seeds to the surface to germinate. Meat produced in this way for local distribution is high in quality and value, with negligible rearing or maintenance costs. It requires little outlay beyond the trees themselves, and can be done in part or by stages contributing to valuable diversity.

Bibliography

Arbury, J. and Pinhey, S. (2002) *Plums*. Beaumont Press, Singapore.

Briggs, S. (2012) *Agroforestry: A New Approach to Increasing Farm Production*. A Nuffield Farming Scholarships Trust report sponsored by the NFU Mutual Charitable Trust. Available at: https://www.nuffieldscholar.org/sites/default/files/reports/2011_UK_Stephen-Briggs_Agroforestry-A-New-Approach-To-Increasing-Farm-Production.pdf (accessed 7 December 2021).

Darwin, C. (1881) *The Formation of Vegetable Mould through the Action of Worms*. John Murray, London.

Game, M. (1999) *In a Nutshell: The Story of Kentish Cobnuts*. Chris Howkins, Surrey, UK.

Knight, S. (2020) Can farming make a space for nature? *The New Yorker*, online. Available at: www.newyorker.com/magazine/2020/02/17/can-farming-make-space-for-nature (accessed 29 September 2021).

Shepard, M. (2013) *Restoration Agriculture: Real-World Permaculture for Farmers*. Acres USA, Austin, Texas.

Smith, J.R. (2015; first published 1929) *Tree Crops: A Permanent Agriculture*. Midwest Journal Press, Mexico, MO.

The Woodland Trust. Available at: www.woodlandtrust.org.uk/ (accessed 29 September 2021).

Wohlleben, P. (2017) *The Hidden Life of Trees*. William Collins, London.

11 Case Studies

Case Study 1: Becklands Farm, Marshwood Vale, Dorset, UK

Becklands Farm is 10.5 ha, which is currently mostly pasture land. It is owned and farmed by one person, Hilary, a widow in her 70s, the farm having been her husband's family farm. The farmer, Francis Joyce, a Cambridge agriculture graduate much troubled by asthma, thought that a less toxic environment would improve his health. He sang in a choir with Rolf Gardiner and Lady Balfour, who were founder members of the Soil Association, defining lasting standards of organic farming. The farm went organic in 1972 and the farmer had no further recurrence of asthma.

Taking on the farm in 1995, Hilary and her husband kept 300 chickens, feeding them on organic feed or their own home-grown barley, which was acceptable as a conversion feed. The farm was recertified in 1997 and joined the Defra Countryside Stewardship Scheme, which was particularly interested in the Marshwood Vale and its old poplar trees.

The Stewardship Scheme provides financial incentives for farmers and land managers to help their environment by conserving and restoring wildlife habitats, managing flood risk, creating and managing woodland, reducing agricultural pollution and preserving historical landscape features. The scheme is open to all land managers and land-use types and is competitively scored against local priority targets.

The addition of more land and diversification in the form of a holiday cottage achieved viability for the farm. The Stewardship Scheme gave grants towards fencing and in one case the preservation of a fallen poplar tree. Hilary thought this strange at first until she saw how the sheep loved it and the insect life that it harboured. In 2011, they sowed herbal lays on two-thirds of two fields. The continuity of the permanent pasture has held up well, and even after 9 years, the animals love it and will always graze there first. The health of the animals has also been good on this grazing. Only twice has Hilary had to use chemicals on her land. During the foot-and-mouth epidemic, she was obliged to use a formaldehyde footbath in the gateways, and she once got special permission from the Soil Association to use CLiK, a remedy for a parasitic fly on the sheep. The treated sheep could not be sold for meat for the following 80 days, but if a herbal remedy is used instead, they can be marketed at the optimum time. The sheep are Dorset Polls encouraged by the Stewardship Scheme who favour local breeds where possible as they are better suited to the conditions. Dorset Polls are said to cause less soil compaction. Wondering if her sheep should be dewormed, she had their faeces tested and found they were clear of infestation. Hilary put this down to self-medication on chosen plants within the variety available. Indeed, plants containing tannins would have this effect. Noticing that the animals ate ivy when they could, she now offers it to them.

The farm has also had little trouble with weeds. After 2 years of cutting nettles and docks, an insect arrived that devoured all the field weeds in a short space of time, and they have not reasserted themselves.

DOI: 10.1079/9781789243604.0011

It is tough going making a living on so few hectares, a maximum of 30 sheep being suitable for the land. Although not profitable in themselves, Hilary saves on transport and medications. She has the additional premium of the organic label for her produce, and by selling directly to customers by email, she makes the most that she can.

Case Study 2: Mead Farm, Dorset, UK

Mead Farm is some 607 ha of which 324 ha are arable and 283 ha are permanent pasture (seeded about 20 years ago), as well as 81 ha of scrub and woodland.

The fields are between 4 and 8 ha, divided by hedges. The soil is a fertile light sandy loam known locally as Bridport sand. The land has been in the family for 100 years, but for the declining years of the previous incumbent had been farmed by contractors with the inevitable toll on fertility. The extensive acreage of winter wheat and maize left the soil bare for much of the winter, with soil erosion in the wet winter of 2019/20 painfully evident in the deep runnels and ponds of fine silt on the downhill edges of the fields (Figure 11.1).

The current owner, James, is keen to redress this and has set out his list of concerns over the difficulties of changing direction away from chemical farming. They are: (i) weeds; (ii) disease; (iii) yields/margins; (iv) developing new markets; and (v) complexity/ change (the need to learn and devise new systems and tricks of the trade).

The main concern when abandoning herbicides is always weeds. I had had no intention of coming up with any pat answers during the interview, but it seemed that James and his farm manager, Giles, were so keen to find some answers that it was inevitable to discuss some options. Pointing out that hoeing and tilling were just what every weed seed

Fig. 11.1. Typical soil degradation by erosion after being exposed for a wet winter.

loves, because it turns the seed under for protection from birds and creates a nice bare space into which to germinate, sounded like common sense. The next step would be to find an undersowing candidate for maize in particular that would protect the soil, preferably add nitrates to it and not inhibit the harvesting procedures. A low-growing legume would do this. One of the weeds that was seen as troublesome was the volunteer barley that grew among the turnips. It did not look like good farming practice to have it there and it inhibited the early growth of the turnips, having fallen through the sieves during harvest. If turnips followed wheat, this did not happen.

Disease is another main concern. Again, it was accepted as common sense that pathogens will accumulate most quickly and ferociously where there is an abundance of prey. Monocrops provide this abundance without competition. The further cultivation can move towards diversity, the less troublesome the pathogens will be. The healthier the soil is in terms of microorganisms, the better the balance of prey and predators, and the less likely that a pest or disease will get out of hand. In the meantime, while waiting for this perfect state to arrive, the minimum of chemical applications, with very careful timing, would achieve protection with the minimum of harm to the ecological balance. A possibility for future consideration might be using a companion plant for crops that suffer from attacks from aphids. Buckwheat, which hosts aphid predators and also secretes chemicals that attract phosphorous-producing organisms, and pollinators could be choice candidates.

Profit margins are a recurring anxiety to farmers as their costs escalate. Mead Farm has devised a rotation that relies deliberately on farmyard manure. They grow maize for livestock farmers in return for their manure, which is brought to spread on the fields before planting. The livestock farmers get the maize from the fields where they have put the manure. Nevertheless, some fertilizers are also bought in for other crops. It is hoped that a greater expenditure on fertilizers is rewarded by a greater yield, but as soils become increasingly sterile, this is definitely not a given. Promoters of agrichemicals encourage the mindset of equating success with yield, but profit margins are helped equally by reduced expenditure. At Mead Farm, herbicide for 121.4 ha at £160.60 ha^{-1} costs £19,500. Additional maize starch for sowing with the seed costs £295 t^{-1}. During the year, £41,000 had been spent on fertilizers and £39,000 on sprays. Saving on these expenses while maintaining yield would be a worthwhile target to improve the profit margin. No insecticide has been used for the last 5 years, and there has been no appreciable improvement in aphid infestations. Seed dressings were used until they were banned and it was found that untreated seed was eaten by birds. The barley yellow dwarf virus has become a problem, getting established on young crops that are sown early and curtailing the yield. Delaying sowing to avoid this has had unwelcome consequences.

If winter set in early, large areas could not be sown at all with winter wheat, and erosion of the bare soil in a wet winter was bad. It will now be sown with more maize in the spring so there is no rotation. This misfortune is attributed to loss of seed dressing, but also highlights James's concern that they 'have to learn new systems and tricks of the trade'. These are seen as complex and requiring a whole new approach. As barley yellow dwarf virus is spread by aphids, aphid predators such as ladybirds should be encouraged, and fitting buckwheat into a rotation would help this, as will James's resolve to encourage wildlife.

James's concerns with continuing with his current regime are: (i) wildlife destruction in a wide sense; (ii) erosion; (iii) a very weak negotiating position when buying and selling (we have 'perfect competition' between farmers, but all the people we deal with are oligopolies) ; and hence (iv) low overall levels of profit.

The problems that Mead Farm would like to resolve first are the loss of wildlife and soil erosion. A good start is the project to sow strips of wild flowers on the field margins, and to use cover crops to protect the soil, which can simultaneously provide the diversity for wild life and soil organisms. Seed is much cheaper than chemical additives. Interplanting and undersowing will take more careful planning and timing, but as the land responds, further savings can be made.

Mead Farm is not alone in finding themselves at the mercy of the few large organizations that take their produce. They would love to find new markets. While small farms can often sell direct to the public and increase their margins that way, it is not so easy for a large farm to do that. Gabe Brown, a farmer in North America, has resolved this problem by licensing other operators to use portions of his land for a variety of products that can be marketed locally from a farm shop and to a regular customer base. By offering farm work experience to agricultural students, he has developed a pool of reliable tenants who are happy to start their farming careers in this way. In addition to direct marketing, his land benefits from the diversity of activities on it, which improve the ecology.

Mead Farm has already diversified and built a 'village' with a cafe, events venue and rural traders. This now lacks only a farm shop and the supplies for it.

James is keen to embrace a new direction that will move him into profitable and sound ecological land management. The attraction of change is to have much healthier land and more wildlife, plus healthier food and the ability to sell better food direct to customers and not be caught in the pincer movement between buyer and supplier. Working with nature rather than against it needs a greater awareness and understanding of it, but this is also enjoyable and rewarding.

In May 2021, we received the following email from James: 'Your advice has been critical. We are in the process now of changing to regenerative agriculture and have done trials on planting two fields of maize using a Claydon hybrid M4 drill and it looks very promising – much less erosion.'

Thanks to James and Giles for agreeing to take part in this case study. I think it will have a resonance with many other farmers in a similar position, and I hope that their participation will have immediate rewards for them.

Case Study 3: Meg Tebbs, West Dorset, UK (no-dig vegetable patch)

Deep cultivation can cause considerable damage to soil structure, disturbing the microorganisms that live in and maintain it. This damage applies to domestic vegetable gardens as well as to farmland, where a spade is used instead of a plough. The no-dig method of cultivation has been around for a while but is still not completely accepted. I decided to set up a no-dig vegetable garden, in order to observe whether there are any benefits to this method on a small scale. After years of traditional digging, this would be a voyage of discovery. Received gardening wisdom is to turn the soil over to a spit deep (spade depth and sometimes even double that), to remove all weeds and to create a fine tilth, suitable for sowing seeds. This can result in subsoil being brought to the surface, especially in heavy clay areas where it can take some time to break down.

Traditional digging has several drawbacks: (i) all the weed seeds buried in the soil are brought to the surface by digging and exposed to light, where they germinate rapidly; (ii) worms are disturbed and often chopped up; (iii) soil microorganisms are churned up and displaced; and (iv) this sort of digging can be hard, back-breaking work.

Many would-be gardeners have been dismayed by the rapid way weeds germinate and grow, rapidly infesting their carefully cultivated area. This can happen while they take a few days off to recover from digging.

In order to try out and evaluate the no-dig way, some spare ground in my garden was selected, an oblong piece of land measuring about 4 × 1.5 m. This size is not too big for most gardens, and is narrow enough to be able to reach into the middle from both sides. The grass was roughly cut back, and any weeds with deep taproots, such as dandelions and docks, were dug up. Bindweed (*Calystegia sepium*) and field bindweed (*Convolvulus arvensis*) are almost impossible to remove, as their brittle roots can reach several feet underground. Covering and omitting light should weaken growth to manageable levels, and occasional hoeing will gradually eliminate it.

The cleared ground was then completely covered with cardboard (e.g. old boxes) and old bits of carpet (without rubber backing) (Fig. 11.2). This is a tried-and-tested way of eliminating light and is intended to kill off any unwanted plants. Over time, the cardboard will rot into the soil, providing extra nutrients, and the carpet can be reused.

A rough raised bed was made by nailing together four pieces of old planking that were lying around in the garden (Fig. 11.3). The planks were about 15 cm deep, enough to contain compost.

Fig. 11.2. Rough ground covered in cardboard and old carpet.

Fig. 11.3. A rough raised bed with deep layer of added compost.

The compost was ordered in bulk from a local organic farmer, made using well-rotted manure from his own beef herd, combined with lots of green and woody plant materials. The manure undergoes a high-temperature aerobic composting process to deodorize and sterilize the material. It is then left to stabilize and mature before being blended with the other organic materials. This process takes several weeks. The farmer does not use pesticides or herbicides (e.g. aminopyralids; see Appendix), so his product could be used with confidence. Many gardeners will have their own home-made compost, and I plan to use my own in future, when enough has been produced.

The compost was tipped into the bed to a depth of about 5 cm and raked level. Cultivation was ready to commence, once the soil had warmed.

Vegetable seeds came from a variety of sources – some from old left-over collections and some from online seed companies. Onion and parsnip seeds do not last more than a year, and so fresh seed was obtained. Onion seeds were sown inside in the warmth, in January or February, to allow as much time as possible to make good bulbs.

After germination, the seedlings were transferred to small pots or modules to continue growing in a glasshouse.

In March, beetroots, spring cabbage and lettuce were started in small modules under cover, to plant out once the ground had warmed up. Parsnip and carrots do not transplant well, and therefore were sown directly into the compost once the soil had warmed up.

In April, with spring well advanced, the raised bed was planted up. Beetroots, lettuce, onions and spring cabbage were put in and left to grow (Fig. 11.4). Space was left at one end for two courgette plants, which were sowed in May after the last frost. The plot was watered once or twice a week, if no rain occurred. The deep fibrous compost held moisture well, and even when there were a few weeks of drought, no wilting of the crops occurred.

The row of spring cabbages (heritage variety Greyhound) grew large and developed good hearts. They needed to be covered with netting to keep cabbage white butterflies away, whose caterpillars can quickly devastate a crop. The cabbages were ready to harvest from late June onwards, as were the lettuces. These were a crisp Iceberg variety and Lollo Rosso (a cut-and-come-again type) planted along the length of the bed. Beetroots, both purple and golden, were harvested when they reached golf-ball size. The courgettes that were planted in late May started producing fruit in mid-July. Courgettes need plenty of rich soil and water to grow well. The two plants thrived and reached a good size in the raised bed.

Carrots sown directly into the compost in April did well. Germination can be erratic, and seed was sown every few weeks. More vegetables were started in modules, ready to put out as older plants were harvested. The raised bed was always in use, and small amounts of compost were used to replenish the empty spaces. Fennel and spinach, which grows best after mid-summer, were sown in late July, ready to go out in August. Salad crops were regularly sown, and winter crops were started in modules in spring, such as Brussels sprouts and leeks, and were planted between onions and carrots. Crops sown in August and September when the ground is still warm can put on enough growth to

Fig. 11.4. Cabbages, lettuce, onions, beetroots and courgettes.

survive the winter months, especially with some cover from wind and heavy rain. These crops were grown under cloches or fleece, and harvested through winter as needed.

The compost in the raised bed proved to be very fertile, and harvests exceeded all my expectations. The soil structure was improved as the compost was gradually drawn down by worms. Any weeds that germinated were easily removed from the deep compost when small, saving a lot of time. The amount of vegetables raised from a comparatively small area was remarkable. This system allows cropping to be seasonal and provides harvests for most of the year. The compost does not need to be replenished for at least a year, instead just adding a handful when planting a second crop. The bed is easily covered by netting or cloches to protect plants from pests, birds and bad weather. No pesticides were used, just nets and vigilance. This proves that anyone with a small amount of space can adapt this method to their needs and circumstances to grow their own food crops. A small area of soil outside the raised bed was sown with meadow flowers such as ox-eye daisies, sweet Williams, cornflowers and rudbeckia (Fig. 11.5). These attracted plenty of bees, hoverflies and other insects good for pollinating crops such as beans and courgettes. Harvesting of lettuces began in May, and beetroots and also some cabbages in June. The red onions were rather small, which may have been due to the variety, as onions in another part of the plot grew large. It was good to plant a second batch of vegetables as soon as space became free, topping up with some compost. Cauliflowers were grown successfully for the first time and, with netting to keep out cabbage white butterflies, grew to a good size. The bed will continue to be planted up through autumn, with cool-temperature plants such as pak choi, spinach and lettuces. This will allow enough growth before the short days and lower temperatures slow things down. This experiment has

Fig. 11.5. Pollinator-friendly planting in the surrounding area.

shown how much can be cultivated in a small area, without too much effort and with some forward planning.

Case Study 4: Tim Parton, Farmer, Staffordshire, UK

The farm is 300 ha of arable land. The soil type is sand and clay. Tim started soil improvement on the farm about 10 years ago, after becoming aware of the damage caused by repeated use of chemical fertilizers, pesticides and herbicides. The soil structure was lacking carbon and nitrogen, and earthworms were few to non-existent. He switched to strip tillage, planting a diverse mix of cover plants and gradually improving the biology of the soil. After about 3 years, the worms had returned, and the soil fertility had improved greatly. Since then, Tim has eliminated insecticides and seed dressings, and is reducing the use of herbicides and fungicides. His modified drill enables him to plant several crops at one time, along with some of his home-made biological brew and the cover crops. Using thick rooted oil radish (*Raphanus raphanistrum*) improves drainage. Berseem (Egyptian) clover (*Trifolium alexandrinum*) is an aggressive cover plant through the summer, before succumbing to frosts. The eventual aim is to establish a hardy white clover (*Trifolium repens*), which could remain on the surface, suppressing weeds and covering the soil at all times. Legumes including lupins add nitrogen to the soil, and Tim grows these as a cash crop. The legumes are grazed by sheep, and their manure adds to the fertility. In freezing weather, Tim will go out about 3 a.m. and roller any legume remains so that they shatter. Worms will eventually draw the broken bits down into the soil.

Tim makes his own biological brew in a 1000-litre tank, using Aiva BioPlus powder, which he says plays a vital role in fixing nitrogen from the atmosphere, releasing phosphorus from the soil and fighting fungal infections on the leaves. Molasses are added to nitrogen applications to reduce the amount needed for each crop and to feed the soil bacteria and fungi. In the first year of using the brew, he was able to cut applied nitrogen by 40 kg ha^{-1}, while still getting a good yield. He says that this is only possible because of his good soil. By putting soil biology first, Tim's drains run clear, his water infiltration rates are excellent, and the soils are spongy and deep. Crops can be produced with fewer or no chemicals. Tim plans to rely on biology in the future, but will keep chemicals near, just in case! However, he has not used insecticides for 4 years, pointing out that they are inefficient and wipe out useful predators.

The use of glyphosate is restricted, and there is no spraying in autumn. A deep compost mulch on top of grass weed is enough to kill it. Yields have gone up. The main crops – wheat, millet, malting barley, beans, oats and lupins – are doing well. Wheat yields alone have improved hugely, and the farm now saves about £50,000 a year compared with 10 years ago (Fig. 11.6).

The farm grows a wide rotation, which includes oilseed rape with a companion crop every 6 years. Tim has also started a hay business to serve the local equine community. This allows him to keep grass in the rotation, putting carbon back in the soil and boosting mycorrhizae. Cover crops are working well, and fertility is improving all the time. He uses a compost extract composed of horse manure and plants, which works well.

The soil's increased water absorption and retention is another success, he says. 'When it rains, the infiltration rate – even after a lot of water – is great. It just keeps absorbing and holding that water. Wildlife on the farm has also benefited, with a mix of birds including lapwings, skylarks and greenfinches, along with the top predator, peregrine falcons.

Fig. 11.6. Tim's combine harvester. Copyright © Tim Parton.

Tim is passionate about his journey, learning more and more about sustainable agriculture without damaging the soil and the environment. Wildlife is returning to the farm, including skylarks and golden plover. Insects are back in force. He is always willing to share his knowledge, but accepts that his kinder way of farming may take time to be accepted. The hold of the agrichemicals business is hard to shake, and rushing people into change may scare them off. However, the savings are real, and the rewards are many. The farm is sustainable and is working as it should. Tim was made Arable Innovator of the Year for 2019 and is rightly proud of his achievements.

Case Study 5: Mount Benger Farm, Otago, New Zealand

Mount Benger Farm comprises 2675 ha of grassland and is one of three jointly owned by two families, the Frasers and the Sheldons. Father and son, Duncan and Stuart Fraser, and Stuart's wife, Jo, do the main work on the farm, while other family members do skilled work on machinery and buildings. John, father of the Sheldon family, is instrumental in planning and commercial decisions.

The land has been farmed by the Fraser family since 1922. A near-disastrous experiment in running a larger farm from 2008 to 2015 convinced the family that medium-sized farms did not work, and they decided to reduce the farm size.

Ten years ago, Stuart's parents decided to move off the farm and give someone else an opportunity to farm. He says they still love to be on the farm and involved, without having to make the day-to-day decisions.

'We had a few spare houses on the farm so created a shareholders' house. It's a house they can come to on the weekend and fix the fences that I don't have time to do or shift the irrigators. They love it.'

Stuart has nieces and nephews too, and he says it is nice to give that next generation an insight into what is happening on the farm and in the industry, along with the different opportunities. He says compliance is a big challenge for Mount Benger Farm, with the irrigation scheme along with the typical North Canterbury climate (Fig. 11.7): 'Trying to reduce the amount of nitrogen leaching, trying to be sustainable by fencing off waterways, applying fertilizer at better times of the year, keeping cattle off the flats where possible to reduce pugging, as well as direct drilling more so than conventional cultivation to keep the soil structure intact. But we love it. I think that it's an awesome industry to be in and we are making a lot of progress both as an industry and a farming outfit.'

The livestock load on the land is 2.7 sheep ha^{-1} with an additional 1.5 to two lambs per ewe in the summer. Cows graze at 1.8 cows ha^{-1} with an additional calf in the summer. Mob grazing is rotated so that grasslands can recover between grazing, and strip grazing is used in the winter for cattle being prepared for sale.

The land is seldom ploughed, with 99% of sowing being done by direct drilling to minimize soil disturbance. Oversowing is done occasionally to get more variety in the sward, by mixing seed with an aerially distributed fertilizer. Mixtures of grasses and legumes tend to revert to a dominant species as a result of preferred grazing or shading out by stronger species.

The 2021 floods tested the regime. Stuart said, 'We had 130 mm more than we have had all year but things have soaked in well with no major damage.'

As both families have strongly scientific backgrounds, innovation is a constant theme. John has been in the aerospace business, so new technology is a core feature of any changes. Recently, fertility mapping from the air has allowed fertilizer to be applied exactly where needed and in the exact quantities required so that there is no waste and no leaching. John has written a paper on 'Land use for food production' (see Appendix 2), which addresses the problems of carbon emissions from farming. He concludes that permanent pasture is not a problem but belching ruminants are. Among possible solutions may be treating ruminants with probiotics, inoculating them with a low-methane microbiome, or breeding cows that burp less. Certain pasture mixes, such as those containing docks and sweet clover, also cause fewer methane emissions from ruminants. Non-intensive, grass-fed, meat animals that occupy land unsuited to cereal crops is the aim on this farm. With ever an eye to the future and a preparedness to change, the Frasers and Sheldons would like to meet demand for quality meat while acknowledging that feeding the world in the future will be dependent largely on plant-based foods.

Fig. 11.7. Rolling tussock country on North Canterbury's Mount Benger station. Copyright © Stuart Fraser.

Appendix

Agricultural Chemicals (Agrichemicals)

The meteoric rise of man-made fertilizers, herbicides and pesticides has had long-term consequences not foreseen in the heady days after the Second World War, when agriculture became a high-tech business. Agricultural chemicals are manufactured pesticides, herbicides or fertilizers used for the management of ecosystems in agricultural sectors.

The production of synthetic fertilizers made an entrance at the end of the nineteenth century, and paved the way for modern agricultural production. Crop yields were greatly increased by the combinations of chemicals and inorganic substances, making vital nutrients instantly available to plants. Organic fertilizers include green manures, manures from animals and compost from degraded vegetation, and are released slowly. Man-made fertilizers are easy and convenient to use, and are available to buy from garden centres and retailers. However, beneficial microorganisms in the soil are adversely affected by these products, and excess nitrogen and phosphates can leach into water courses, disrupting aquatic life and causing pollution.

Fungicides

Fungicides target fungi, and disrupt mycorrhizae in the soil.

Herbicides

Herbicides are designed specifically to kill weeds – some will kill only grasses while others will affect other species. Aminopyralids and clopyralids are meant to improve grazing grasses while killing other species of plants. They are so effective that they can pass through an animal's gut and still cause plant damage by lingering in manure and compost, often for several years. This has caused great concern from gardeners and horticulturists, who need to know that their growing mediums are safe.

Glyphosate is perhaps the best known and most controversial herbicide today, along with 2,4-dichlorophenoxyacetic acid (2,4-D), atrazine and dicamba. Glyphosate is absorbed through foliage and minimally through roots, and goes to growing points (meristems). It inhibits a plant enzyme involved in the synthesis of three aromatic amino acids: tyrosine, tryptophan and phenylalanine. It is therefore effective only on actively growing plants. It is used to quickly 'kill off' green manures before planting cash crops. Atrazine is a widespread pollutant of groundwater and drinking water, and has been linked to increased risk of cancer and reproductive problems. It is banned in Europe and many other countries, and is mostly used in the USA. Crops genetically engineered to be resistant to certain herbicides allow the herbicide to coat both crops and weeds, killing the weeds but leaving the crop alive. Each year, the USA uses more than 0.5 billion kg of pesticides. In 2017 and 2018, the Environmental Protection Agency approved over 100 pesticides containing ingredients widely considered to be dangerous. Some of these are considered possible carcinogens by the World Health Organization's International Agency for Research on Cancer. Herbicides like glyphosate and 2,4-D are also common ingredients in non-organic lawn care products.

Asulam is used to control bracken, a toxic fern of acid ground. Asulam is moderately toxic to birds, insects and aquatic life, and its use is controlled.

Parasiticides

Parasiticides are designed to keep parasites at bay. Some are more common in agricultural contexts than others, but all fall under the general pesticide umbrella.

Pesticides

Pesticides target pests that attach to crops but also kill beneficial insects such as bees, hoverflies and beetles. Chlorpyrifos, an organophosphate pesticide that can permanently damage the developing brains of children, was first developed by the Nazis as part of chemical warfare. Widely used in the USA on food crops, it has now, after many years of legal fights, been banned by the US Environmental Protection Agency (www.nytimes.com/2021/08/18/climate/pesticides-epa-chlorpyrifos.html; accessed 30 September 2021).

Neonicotinoids are nerve agents targeting aphids that may infect crops with virus infections. Widely used since the 1980s and 1990s, neonicotinoids are extremely toxic insecticides in infinitesimal doses. In spite of their danger to the environment and living species, which were documented as soon as they were put on the market, it took two decades before the first prohibition measures appeared, particularly in the European Union, and they continue to be used massively on all continents, generating legal action, both by environmental associations and beekeepers to prevent their use, and by agrochemical companies to challenge their prohibition or the limits on their use. Neonicotinoids contain a systemic chemical which is water soluble and may leach into watercourses. This can be absorbed by wild flowers and released in their nectar and pollen. Neonicotinoids have also been shown to cause a wide range of damage to individual bees, such as memory loss and reduction of queen numbers. Evidence has strengthened recently of neonicotinoid damage to colonies of bees, and revealed a huge reduction in all flying insects.

Biodynamic Agriculture

Beset by charges of pseudoscience, and lack of open-minded research, biodynamic agriculture per se has gained little mainstream credibility in the last 100 years. In fact, it operates on mainly organic principles but with elements of resonance and the influence of cosmic rhythms incorporated in the practices. Originally, it started with filling a cow horn with cow dung and burying it for a year. Then a weak solution was made by stirring a handful of the resulting friable contents into three gallons of rainwater in a 5-gallon bucket. Stirring rhythmically in one direction with a long stick and then reversing the direction to create a vortex was key to the procedure, which seemed to charge the water with vital properties. Such water sprayed on the soil gave it noticeable fertility. In due course, because of the difficulties of large-scale production, the water prepared on homeopathic principles became available for spraying commercially. There is plenty of case study evidence of success with this approach. It is tightly bound with organic principles, which have gained mainstream credence, but the notion of harnessing cosmic energy is still outside current acceptable areas of research.

Hügelkultur

Hügelkultur (hügel means hill) involves making a raised, two-sided bed using a base of branches, leaves, grass and straw. This method was adapted to make a raised bed, more suitable for a smaller garden. Sides were made from old planks of wood, and the base was covered in cardboard and plant materials. These were topped with compost or soil.

This is a great way of utilizing garden waste, rather than sending it for kerbside collection. The branches at the bottom decay gradually, providing nutrients for many years, and the herbaceous waste rots down more quickly. The deep layer of compost provides a good habitat for soil microbes, worms and fungi, and moisture is retained (See figs A.1., A.2. and A.3.).

Fig. A.1. First stage with branches over cardboard.

Fig. A.2. Branches covered with garden waste layer and more cardboard, topped with compost.

Fig. A.3. The bed is planted with celery, beetroot and carrots, which are growing vigorously.

Uses of Land for Food Production, by John Sheldon (12 November 2019)

John is part owner of three sheep, cattle and dairy farms in New Zealand. His main business has been in aerospace so that opinions and facts here are gleaned from discussions with his farm's New Zealand fellow directors and from interviews with various people with agricultural expertise in New Zealand.

Introduction

According to scientists the earth has to produce enough nutritious food for up to a maximum 10 billion people. Today the total is 7.75 billion. From January 2019 the number of babies born in the world is declining. The reasons for further increases in population are entirely due to greater longevity, a factor which has many implications.

Scientists also maintain that to feed 10 billion, everyone will have to eat only crops and forego all meat. They admit, however, that this is based just on the fact that it takes 5 kilos of feed to produce 1 kilo of meat and that they have not fully investigated effects of the nitrogen cycle, the effects of phosphate-based fertiliser and available quantities of phosphorous, the relative importance of originators of methane and atmospheric carbon concentration.

This brief looks into these issues as part of the debate about producing animals for meat and milk.

Its conclusion is that extensively farmed sheep and cattle, far from taking crops that would otherwise be consumed by people, actually as a by-product increases the volume and quality of crops, decreases the quantity of phosphorous required to a sustainable level and decreases the carbon concentration in the atmosphere.

Burping by ruminating animals is the third largest natural contributor to methane in the atmosphere and this is undoubtedly a disadvantage. However, it is a problem that may more easily be corrected than other sources of methane.

Many current methods of meat, milk and crop production have bad effects on the nitrogen cycle. Solutions are required to prevent nitrous oxide poisoning rivers and water tables. If anything, the solution applicable to meat production is the easiest to apply.

What needs to go is sheep and cattle raised on feed other than forms of grass that people cannot consume. We need to get rid of animals raised intensively.

There is a very strong case for people eating less but still a reasonable quantity of sheep and cattle meat. Ironically there is a less good case for eating chicken and other farmed birds. There is a modest case for pigs so long as they are fed on food which would otherwise be wasted.

This paper does not deal with what is an ideal diet for people. People should probably be free to choose their diets in the light of what dieticians feel is good for them at various stages of their lives.

Compromises are often the best way out of problems – getting people to eat less meat but not to forbid eating it seems like a sensible policy all round, free from ideological nonsense. This would be especially the case if it is proved that a modest amount of meat eating will engender maximum safe food production.

Governments have a long history of pushing people down roads that are supposed to be good for the environment only to discover that these roads lead only to outcomes that are worse than what they replaced. Cautious progress is almost always better than making massive changes.

Feed for animals

It is accepted by most experts that it takes about 5 kgs of cereal crops to produce 1 kg of sheep or beef meat. In a world that does not have a surplus of food it is clearly wasteful to diminish the amount of food available to humans in this way where it can be avoided.

It is wrong, however, to derive from this fact the conclusion that all production of such meat should cease.

There are many areas that can sustain cattle and sheep that, because of topography and soil types, are unsuited to growing cereals.

Moreover, there is a strong case for periodically rotating animals round cropping fields. The rotation of animals and crops led to big increases in the productivity of the soil in the 18th century and the case for it remains strong today despite developments (or even perhaps because of developments) in fertilisers.

In any event, greater efficiency in breeding, animal selection and parasite control is increasing the yield of meat from cattle and sheep per quantity of grass or other feed consumed.

Phosphorus

Phosphorus is in plentiful supply but it is a diminishing resource as demand exceeds natural generation of supply.

Everything we eat removes phosphate from the soil. Except for the excrement of a falling number of Chinese peasants, this phosphate is largely flushed into the oceans. Eating cereals removes 10 times as much phosphate as eating the same nutritional value of meat. Drinking milk removes 5 times as much.

Phosphate is put back into the soil either by spreading fertiliser or by allowing animals to leave their urine and excrement on fields (either on hill pastures unsuitable for cropping or on cropping fields as part of a rotation).

It is common in countries where animals are intensively farmed in sheds or in limited areas where their main diet is cereal feed for their dung to be spread on cropping fields. This is a good use of the dung but a very poor use of the cereal feed.

Some scientists believe that artificial fertilisation of fields, including large amounts of phosphate has not increased crop yields over the last 70 years, whereas production has increased in farms where cattle and sheep have roamed.

Thus it can also be said that from a soil fertility point of view, rearing cattle and sheep on areas unsuitable for cropping is good, as is rotating them periodically through cropping fields.

Hyperspectral imaging and computerised selective spraying from aircraft is a development that means that phosphate and other fertilisers can be intelligently spread to even out levels whilst avoiding adding levels to soil that does not need it. This leads to a reduction of the use of phosphate especially in hill country farms.

Carbon capture

Grasses grown to feed animals on country unsuitable for crop growing such as hill country in Wales or New Zealand are highly photosynthetic, turning carbon dioxide into oxygen.

These grasses include rye, clover, lucerne and alfalfa. In much of this country, these grasses would not exist, or exist in such quantity and richness, if it was not for the farming of sheep and cattle.

Many cereal crops are far less or not at all photosynthetic as they mature. There seem to be few studies about the relative carbon capture of various crops but no doubt about the above grasses.

Ruminants with their double stomachs feed on grass. Humans cannot eat grass. Such ruminants are therefore not stealing food from humans.

At the time of mad cow disease in Europe, the New Zealand meat marketing board ran advertisements showing cows on green pastures captioned: ‘Cows eat grass, don’t they?’

Methane

Methane eats away at the ozone layer and has other bad effects such as global warming.

Freshwater wetlands are the largest natural producer of methane, followed by rice paddies. Burping from ruminants is also significant. Farting, whether by ruminants or humans, is not a serious cause of methane. Perhaps the most serious increasing source is human use of fossils fuels and general industrial activity.

Researchers in California have been looking at ways of controlling ruminant burping with probiotics. This may be a promising approach since it works with humans.

A recent study of 1,000 cows in various areas of Europe showed that it is possible to breed in animals that emit less methane. An additional approach is to inoculate them with low methane microbiome. No negative side effects were experienced in either of these approaches.

The Nitrogen Cycle

Denitrification, the reduction of nitrates back into nitrogen gas that has been absorbed from the atmosphere into the soil in the first place as part of the essential fertility of the soil is undertaken by various layers of differing bacteria in the soil.

What environmentalists concentrate on is preventing nitrate filtering through the soil into rivers and groundwater.

There has been excessive use of N-fertiliser in agriculture which has led to nitrate pollution in water supplies. This, like the use of phosphates, is increasingly controlled by governments.

Beyond this, the main issue is the application of irrigated water to cropping fields, pastures holding dairy herds and irrigated pastures for sheep and cattle. Through sophisticated moisture measuring and distribution, this problem can be better understood. What remains is unforeseen combinations of irrigation and heavy rains where excessive watering drives nitrates into the groundwater.

Hill country animals and animals on non-irrigated pastures have proved to be the least of the problem here. However, contact between the animals and rivers and streams can stir up a number of different pollutions including nitrate. This can be prevented by appropriate fencing of larger rivers and streams.

Summary

The above notes support the argument that there is a place in the world for meat production from an environmental point of view. Meat eating and vegetarianism should be a matter of individual choice with neither choice being in any sense "better" for food production or the environment. There are nevertheless right and wrong ways of producing meat, milk and cereals.

Having said this, meat eating should perhaps become less prevalent so that the economic drive for bad meat production practices such as shed and cereal feeding is diminished.

Glossary

Actinobacteria: These play an important role in the decomposition of organic materials, such as cellulose and chitin. They are considered high value, as they can produce some antibiotics.
Adventitious root: A root arising from various parts of a plant, such as stems, tubers, bulbs, leaves or the trunk.
Agroforestry: Cultivation of trees in agricultural economy.
Anaerobic: Lacking oxygen.
Angiosperm: A plant that produces seeds that are enclosed in an ovary.
Allelopathy: The suppression of one plant species by another, by releasing toxins.
Anthelmintic: Having activity against parasitic worms.
Bacteroid: An irregularly shaped form of a nitrogen-fixing bacterium (such as *Rhizobium* spp.) found especially in the root nodules of legumes.
Brix tester: A refractometer measuring solids in a solution and correlating them to an established index.
Chitin: Structural polysaccharide made from chains of modified glucose and found in the exoskeletons of insects, the cell walls of fungi, and certain hard structures in invertebrates and fish.
Cordate: Heart-shaped.
Corymb: A flat-topped inflorescence arising from different points up the stem.
Dioecious: A plant with male and female flowers on different plants.
Drench: To forcibly administer a drug in liquid form orally (to an animal).
Ecosystem: The complete system of interaction between plants, animals and environmental factors.
Exudate: Secretion from plant pores.
Halophyte: A plant adapted to saline conditions.
Hermaphrodite: Having male and female parts in one flower.
Layering: Cloning plants by pegging lower stems into the soil so that they root.
Monoecious: A plant with separate male and female flowers on the same plant.
Mucilaginous: Sticky when wet, slimy.
Neonicotinoids: Chemical insecticides related to nicotine.
Nicotine: A chemical produced by *Nicotiana tabacum* (tobacco plant).
Panicle: A much-branched inflorescence, found in grasses such as oat.
Pathogen: An organism that causes disease.
Permaculture: A sustainable and naturally self-sufficient agricultural ecosystem.
Petiole: The stalk of a leaf.
Phosphorus lockout: Occurs when the pH is below a certain level, causing soil to become acidic and preventing root uptake of nutrients.
Pneumatophore: A vertical aerial root with a loose cell structure enabling gaseous exchange for plants growing in mud.

Polyphenolic compound: An antioxidant component found in plant foods; also classified as a micronutrient.

Polysaccharide: A carbohydrate whose molecules consist of a number of sugar molecules bonded together (e.g. starch, cellulose or glycogen).

Raceme: A simple inflorescence, in which the flowers are borne on short stalks at equal distances along an elongated axis, opening in succession towards the apex, e.g. lily of the valley.

Riparian: Living on the banks of rivers or streams.

Sagittate: Spear-shaped

Silvopasture: A system of grazing in woodland.

Subsoil: The layer found immediately below the topsoil; typically less fertile and lighter in colour than topsoil.

Substrate: An underlying substance or layer.

Symbiont: A partner in a close mutually beneficial association.

Systemic: Entering by roots or shoots to the tissues.

Topsoil: The surface layer of soil.

Volunteer: A plant that has not been sown; a 'weed' or previous crop residue.

List of Plant Illustrations in Alphabetical Order of English Names

Index

Note: Page numbers in **bold** type refer to **figures**